Advances in
VIRUS RESEARCH

VOLUME **73**

Advances in
VIRUS RESEARCH

VOLUME **73**

Edited by

KARL MARAMOROSCH
Rutgers University, New Jersey, USA

AARON J. SHATKIN
*Center for Advanced Biotechnology
and Medicine, New Jersey, USA*

FREDERICK A. MURPHY
University of Texas Medical Branch, Texas, USA

AMSTERDAM • BOSTON • HEIDELBERG • LONDON
NEW YORK • OXFORD • PARIS • SAN DIEGO
SAN FRANCISCO • SINGAPORE • SYDNEY • TOKYO
Academic Press is an imprint of Elsevier

Academic Press is an imprint of Elsevier

32 Jamestown Road, London, NW1 7BY, UK
Radarweg 29, PO Box 211, 1000 AE Amsterdam, The Netherlands
30 Corporate Drive, Suite 400, Burlington, MA 01803, USA
525 B Street, Suite 1900, San Diego, CA 92101-4495, USA

First edition 2009

Notice
No responsibility is assumed by the publisher for any injury and/or damage to
persons or property as a matter of products liability, negligence or otherwise,
or from any use or operation of any methods, products, instructions or ideas
contained in the material herein. Because of rapid advances in the medical
sciences, in particular, independent verification of diagnoses and drug
dosages should be made

Library of Congress Cataloging-in-Publication Data

A catalog record for this book is available from the Library of Congress

British Library Cataloguing-in-Publication Data

A catalogue record for this book is available from the British Library

ISBN: 978-0-12-374786-0
ISSN: 0065-3527

For information on all Academic Press publications
visit our website at elsevierdirect.com

Printed and bound in USA
09 10 11 12   10 9 8 7 6 5 4 3 2 1

Working together to grow
libraries in developing countries

www.elsevier.com | www.bookaid.org | www.sabre.org

ELSEVIER   BOOK AID International   Sabre Foundation

CONTENTS

1. **Looking Back in 2009 at the Dawning of Antiviral Therapy Now 50 Years Ago: An Historical Perspective** 1

Erik De Clercq

I. Introduction 3

II. Thiosemicarbazones: The First Antiviral Drugs Found Active Against the Poxvirus Vaccinia Virus 4

III. Renaissance of the Poxvirus Inhibitors: Antiviral Drugs Against a Bioterrorist Poxvirus (Variola Virus) Attack 6

IV. Benzimidazole Derivatives: Second Attempt to Launch the Antiviral Chemotherapy Era 8

V. Renaissance of the Benzimidazole Derivatives: Now Turning into Lead Candidates for the Treatment of Human CMV Infections 11

VI. 5-Substituted 2′-Deoxyuridines: Idoxuridine (IDU) and Trifluridine (TFT), the Third and Definitive Attempt to Unleash Antiviral chemotherapy 13

VII. 5-Substituted 2′-Deoxyuridines: IDU (and TFT) as the Starting Point(s) for Other 5-Substituted 2′-Deoxyuridines, that is, BVDU [(E)-5-(2-Bromovinyl)-2′-Deoxyuridine] 15

VIII. Arabinosyladenine (ara-A), Originally Conceived as an Antitumor Agent, the First Antiviral Drug Licensed and Used for Systemic treatment 16

IX. Acyclovir: The Start of the Selective Antiviral Chemotherapy Era, and Still the "Gold Standard" for HSV Therapy 18

X. Anti-influenza Virus Therapy: A First Attempt (DRB), Followed by a Second (Amantadine) and a Third Attempt (Neuraminidase Inhibitors) 20

XI. Ribavirin and Interferon, Two "Old-Timers", Joining Forces in the Treatment of a Relatively New Disease, Hepatitis C 22

XII. (S)-9-(2,3-Dihydroxypropyl)adenine (DHPA), the First Acyclic Adenosine Analog, Leading to S-Adenosylhomocysteine (SAH) Hydrolase Inhibitors as Broad-Spectrum Antiviral Agents 23

XIII. (S)-9-(2,3-Dihydroxypropyl)adenine (DHPA) Leading to the First Acyclic Nucleoside Phosphonate, (S)-9-(3-hydroxy-2-phosphonylmethoxypropyl)adenine (HPMPA), as a Broad-Spectrum Antiviral Agent 25

XIV. 9-(2-Phosphonylmethoxyethyl)adenine (PMEA), the Sister Compound of HPMPA 26

XV. From PMEA (Adefovir) to PMPA (tenofovir): It All Depends on the Substitution of a Methyl Group for a Hydrogen 28

XVI. Suramin, the First Antiviral Drug Ever Shown to Inhibit HIV Infection
Both *in vitro* and *in vivo*   29

XVII. The Nucleoside Reverse Transcriptase Inhibitors (NRTIs) with
Azidothymidine (AZT) as the Starting Point   30

XVIII. The Non-Nucleoside Reverse Transcriptase Inhibitors (NNRTIs),
with the HEPT and TIBO Derivatives as the Starting Point   32

XIX. The HIV Protease Inhibitors (PIs), Hailed From Their Inception, as
Resulting from rational design   34

XX. New HIV Inhibitors, Targeted at Either Fusion (Enfuvirtide), Coreceptor
usage (maraviroc), or Integrase (Raltegravir)   40

XXI. Conclusion   42

Acknowledgment   43

References   43

**2. Use of Animal Models to Understand the Pandemic Potential
of Highly Pathogenic Avian Influenza Viruses**   **55**

Jessica A. Belser, Kristy J. Szretter, Jacqueline M. Katz, and
Terrence M. Tumpey

I. Introduction   56

II. Influenza A Virus Subtypes and Host Range   57

III. Avian Influenza A Virus in Humans   60

IV. Use of the Mouse Model to Study Influenza Virus Pathogenesis   63

V. Use of the Ferret Model to Study Influenza Virus Pathogenesis   71

VI. Molecular Basis of Avian Influenza Pathogenesis   75

VII. Conclusions   82

Acknowledgments   84

References   84

3. Virus Versus Host Cell Translation: Love and Hate Stories   **99**

Anastassia V. Komarova, Anne-Lise Haenni, and Bertha Cecilia Ramírez

I. Introduction   103

II. Regulation Prior to Translation   104

III. Initiation of Translation   111

IV. Elongation of Translation   137

V. Termination of Translation   141

VI. Conclusions   146

Acknowledgments   147

References   147

Index   171

CHAPTER 1

Looking Back in 2009 at the Dawning of Antiviral Therapy Now 50 Years Ago: An Historical Perspective

Erik De Clercq

Contents

I. Introduction 3
II. Thiosemicarbazones: The First Antiviral Drugs Found Active Against the Poxvirus Vaccinia Virus 4
III. Renaissance of the Poxvirus Inhibitors: Antiviral Drugs Against a Bioterrorist Poxvirus (Variola Virus) Attack 6
IV. Benzimidazole Derivatives: Second Attempt to Launch the Antiviral Chemotherapy Era 8
V. Renaissance of the Benzimidazole Derivatives: Now Turning into Lead Candidates for the Treatment of Human CMV Infections 11
VI. 5-Substituted 2′-Deoxyuridines: Idoxuridine (IDU) and Trifluridine (TFT), the Third and Definitive Attempt to Unleash Antiviral chemotherapy 13
VII. 5-Substituted 2′-Deoxyuridines: IDU (and TFT) as the Starting Point(s) for Other 5-Substituted 2′-Deoxyuridines, that is, BVDU [(E)-5-(2-Bromovinyl)-2′-Deoxyuridine] 15
VIII. Arabinosyladenine (ara-A), Originally Conceived as an Antitumor Agent, the First Antiviral Drug Licensed and Used for Systemic treatment 16

Rega Institute for Medical Research, Department of Microbiology and Immunology, K.U.Leuven, B-3000 Leuven, Belgium

Advances in Virus Research, Volume 73
ISSN 0065-3527, DOI: 10.1016/S0065-3527(09)73001-5

1

IX. Acyclovir: The Start of the Selective Antiviral
Chemotherapy Era, and Still the "Gold Standard"
for HSV Therapy   18
X. Anti-influenza Virus Therapy: A First Attempt
(DRB), Followed by a Second (Amantadine) and
a Third Attempt (Neuraminidase Inhibitors)   20
XI. Ribavirin and Interferon, Two "Old-Timers", Joining
Forces in the Treatment of a Relatively New
Disease, Hepatitis C   22
XII. (S)-9-(2,3-Dihydroxypropyl)adenine (DHPA), the
First Acyclic Adenosine Analog, Leading to
S-Adenosylhomocysteine (SAH) Hydrolase
Inhibitors as Broad-Spectrum Antiviral Agents   23
XIII. (S)-9-(2,3-Dihydroxypropyl)adenine (DHPA) Leading
to the First Acyclic Nucleoside Phosphonate, (S)-9-
(3-hydroxy-2-phosphonylmethoxypropyl)adenine
(HPMPA), as a Broad-Spectrum Antiviral Agent   25
XIV. 9-(2-Phosphonylmethoxyethyl)adenine (PMEA),
the Sister Compound of HPMPA   26
XV. From PMEA (Adefovir) to PMPA (tenofovir): It All
Depends on the Substitution of a Methyl Group
for a Hydrogen   28
XVI. Suramin, the First Antiviral Drug Ever Shown to
Inhibit HIV Infection Both in vitro and in vivo   29
XVII. The Nucleoside Reverse Transcriptase Inhibitors
(NRTIs) with Azidothymidine (AZT) as the Starting
Point   30
XVIII. The Non-Nucleoside Reverse Transcriptase
Inhibitors (NNRTIs), with the HEPT and TIBO
Derivatives as the Starting Point   32
XIX. The HIV Protease Inhibitors (PIs), Hailed From Their
Inception, as Resulting from rational design   34
XX. New HIV Inhibitors, Targeted at Either Fusion
(Enfuvirtide), Coreceptor usage (maraviroc), or
Integrase (Raltegravir)   40
XXI. Conclusion   42
Acknowledgment   43
References   43

Abstract   In 1959, 5-iodo-2'-deoxyuridine (IDU) was described, the first antiviral drug ever to be (and still) marketed (for the topical treatment
of herpetic keratitis). Now 50 years following the description (of
the synthesis) of IDU, we have 50 compounds on the market that
have been licensed for clinical use in the treatment of virus infections. Of those 50, exactly 25 have been formally approved as anti-
HIV drugs; the other 25 have been formally approved for the

treatment of other virus infections: herpes simplex virus (HSV), varicella-zoster virus (VZV), cytomegalovirus (CMV), hepatitis B virus (HBV), hepatitis C virus (HCV), and influenza virus infections.

I. INTRODUCTION

In 2009, it will be 50 years ago that the synthesis of 5-iodo-2′-deoxyuridine (IDU), the first small-molecular-weight antiviral drug to be licensed for clinical use, was reported by Prusoff (1959). Now 50 years after this report appeared, we have about 50 antiviral compounds available in the market, 25 of which are for the treatment of human immunodeficiency virus (HIV) infections and the other 25 are for the treatment of herpesvirus (HSV: herpes simplex virus, VZV: varicella-zoster virus, CMV: cytomegalovirus), hepatitis B virus (HBV) and hepatitis C (HCV), and influenza virus infections.

The year 1929, when Alexander Fleming reported his serendipitous seminal observation that bacterial growth could be arrested by a *Penicillium* species tentatively characterized as *P. rubrum* (Fleming, 1929), is generally considered as the year of birth of the antibiotics era, to be recognized by several Nobel Prizes (for prontosil rubrum as a prelude to the discovery of the sulfonamides, to Gerhard Domagk in 1939; for the discovery of penicillin, to Alexander Fleming, Howard W. Florey, and Ernst B. Chain in 1945; and for the discovery of streptomycin, "the first antibiotic effective against tuberculosis," to Selman A. Waksman in 1952). But, how did the antivirals era start? As I wrote in 1997, viral diseases have, for a long time, been considered intractable by chemotherapeutic means because of the innate association of viruses with the normal cell machinery (De Clercq, 1997).

As early as 1955, Frank L. Horsfall Jr. wrote that the major difficulty has not been to find substances that inhibit virus reproduction ... but to discover substances that will restrict virus multiplication in human beings without at the same time causing damage to the patient (Horsfall, 1955). Eight years earlier, Horsfall Jr. and McCarty had reported that certain substances of bacterial origin (i.e., capsular polysaccharides) could suppress an invariably fatal course of infection with pneumonia virus of mice (PVM) (Horsfall and McCarty, 1947). The most active inhibitors were, oddly enough, the capsular polysaccharides of *Klebsiella pneumoniae*. Of course, *K. pneumoniae* polysaccharides could hardly be considered as "drugable" compounds.

More "druglike" chemicals such as acridine compounds were reported in the early 1950s to inhibit the growth of several viruses (Eaton *et al.*, 1951, 1952; Hurst *et al.*, 1952). Specifically, some amino sulfonic acids were found to inhibit the reproduction of influenza virus in tissues culture, chick embryos, and mice (Ackermann, 1952;

Ackermann and Maassab, 1954). Moore and Friend (1951) described the *in vivo* protective effect of 2,6-diaminopurine on the course of Russian Spring Summer Encephalitis (RSSE) virus infection in the mouse; Jungeblut (1951) described the effect of naphthoquinonimine on infection of mice with the Columbia-SK group of viruses; and Thompson *et al.* (1951a) that of phenoxythiouracils on vaccinia virus infection in mice, but even as early as 1947, Thompson (1947) had pointed to the inhibitory effects of metabolites and antimetabolites on the growth of vaccinia virus in tissue cultures and in 1949, Thompson and his colleagues, including the 1988 Nobel laureates George H. Hitchings and Gertrude ("Trudy") B. Elion, described the inhibitory effects of antimetabolites (such as 2,6-diaminopurine) and α-haloacylamides on vaccinia virus multiplication *in vitro* (Thompson *et al.*, 1949a,b).

II. THIOSEMICARBAZONES: THE FIRST ANTIVIRAL DRUGS FOUND ACTIVE AGAINST THE POXVIRUS VACCINIA VIRUS

That thiosemicarbazones may have potential as chemotherapeutic agents stems originally from the observations of Domagk *et al.* (1946) to whom, according to Bauer (1972), benzaldehyde thiosemicarbazones (Fig. 1) were sent by chance to be examined for *in vitro* activity against *Mycobacterium tuberculosis*. The compounds were found to be highly active against the tubercle bacillus (Domagk *et al.*, 1946). The authors remarked that "das Swefelatom der Thiosemicarbazone spielt eine wichtige Rolle; entsprechende Semicarbazones sind nur gering wirksam" (Domagk *et al.*, 1946).

Because benzaldehyde thiosemicarbazone and its derivatives (i.e., *p*-aminobenzaldehyde thiosemicarbazone) had attracted so much attention in the field of antibacterial chemotherapy, they were the first to be tested, and found to be active, against vaccinia virus in mice and in chick embryos (fertile eggs) for which an assay system had been worked out by Hamre *et al.* (1950, 1951). From a broader structure–function analysis study of the inhibitory effects of the thiosemicarbazones on vaccinia virus infection in mice (Thompson *et al.*, 1953a), it was concluded, not

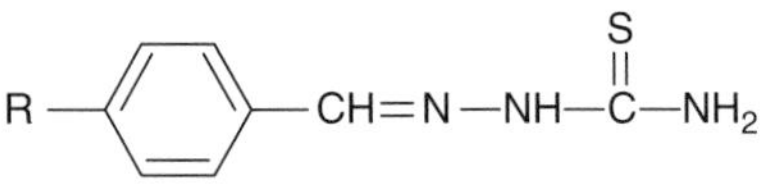

$$R-\langle\text{C}_6\text{H}_4\rangle-CH{=}N{-}NH{-}\overset{\overset{\textstyle S}{\|}}{C}{-}NH_2$$

Benzaldehyde thiosemicarbazone (*R*=H)
p-aminobenzaldehyde thiosemicarbazone (*R*=NH$_2$)

FIGURE 1

only that the $\left[=\text{N}-\text{NH}-\overset{\displaystyle\underset{\|}{\text{S}}}{\text{C}}-\text{NH}_2\right]$ group was essential for antivaccinia virus activity, but that the isatin thiosemicarbazones (Fig. 2) had higher activity and were better tolerated. The benzaldehyde thiosemicarbazones were not further developed, although they could have yielded an anti-viral agent useful in human medicine if their development had not come to a halt (Bauer, 1972).

Both benzaldehyde thiosemicarbazone and isatin thiosemicarbazone were found to protect mice against vaccinia virus infection (Thompson *et al.*, 1951b, 1953b). Isatin 3-thiosemicarbazone even conferred up to 99.99% protection against mortality in mice infected intracerebrally with a neutropic strain of vaccinia virus (Bauer, 1955). In further work a number of derivatives with greater activities were discovered (Bauer and Sadler, 1960), of which *N*-methylisatin β-thiosemicarbazone (Fig. 3) was selected for further study. Although it was not the most active member of the series, practical factors such as ease and cost of preparation gave it some advantage over related compounds. Methisazone has been the subject of a number of clinical trials, that is, in the prophylaxis and treatment of smallpox, and the prophylaxis and treatment of the compli-cations of smallpox vaccination (i.e., eczema vaccinatum, vaccinia gang-renosa) (Bauer, 1972). Reportedly, methisazone would give better results

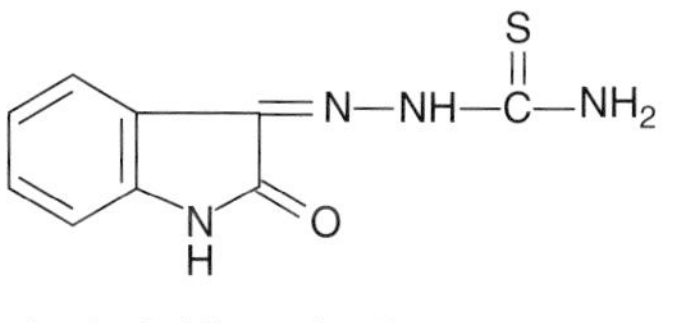

Isatin 3-thiosemicarbazone

FIGURE 2

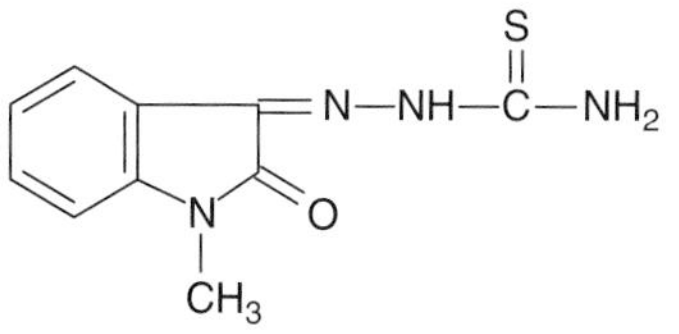

Methisazone
N-methylisatin β-thiosemicarbazone
Marboran®

FIGURE 3

than antivaccinial gamma-globulin in the prophylaxis of smallpox in persons who had been in contact with the variola virus (Bauer *et al.*, 1963).

However, with the imminent global eradication of the variola virus, which in 1980 was officially declared as achieved by the World Health Organization (WHO), interest in the use of methisazone for either the prophylaxis or treatment of smallpox, or the complications of the small-pox vaccination with the vaccinia virus, waned, and the prophylactic or therapeutic use of methisazone, or thiosemicarbazones at large, was no longer pursued.

III. RENAISSANCE OF THE POXVIRUS INHIBITORS: ANTIVIRAL DRUGS AGAINST A BIOTERRORIST POXVIRUS (VARIOLA VIRUS) ATTACK

Although for many years interest in developing safe and effective inhibi-tors of poxvirus infections waned, it waxed again with the since 2000 increasing fear that variola virus, as the causative agent of smallpox, could be used as a potential bioterrorist weapon. In 2001, I reviewed the different classes of compounds that are effective as inhibitors of vaccinia virus multiplication and that, therefore, could be envisaged for the che-motherapy of (ortho) poxvirus infections (De Clercq, 2001). This review was complemented by a list of compounds exhibiting antiorthopoxvirus activity in animal models (Smee and Sidwell, 2003). In this report, the thiosemicarbazones were mentioned as deserving further consideration "because not sufficiently studied in lethal (orthopoxvirus) infection models" (Smee and Sidwell, 2003).

At present, cidofovir (Fig. 4), which has since 1996 been formally approved for the treatment of CMV retinitis in AIDS patients, is still the drug of choice for off-label use in the therapy and short-term prophylaxis of smallpox (should it occur in the context of a bioterrorist attack) and monkeypox as well as for the treatment of complications of vaccinia that could arise in immunosuppressed patients inadvertently inoculated with the smallpox vaccine (De Clercq, 2002). In a murine model mimicking progressive disseminated vaccinia in humans, cidofovir treatment prompted rapid healing and regression of the lesions (Neyts *et al.*, 2004).

The problem with cidofovir and the acyclic nucleoside phosphonates in general is that they have poor, if any, oral bioavailability. To overcome this problem, Karl Hostetler and his colleagues designed alkoxyalkyl [i.e., hexadecyloxypropyl (HDP) and octadecyloxyethyl (ODE)] esters of (*S*)-HPMPC (Fig. 5) and of (*S*)-9-[(3-hydroxy-2-phosphonylmethoxy)-propyl]adenine [(*S*)-HPMPA]], and these oral prodrugs of (*S*)-HPMPC and (*S*)-HPMPA proved, as a rule, highly effective in the oral treatment of

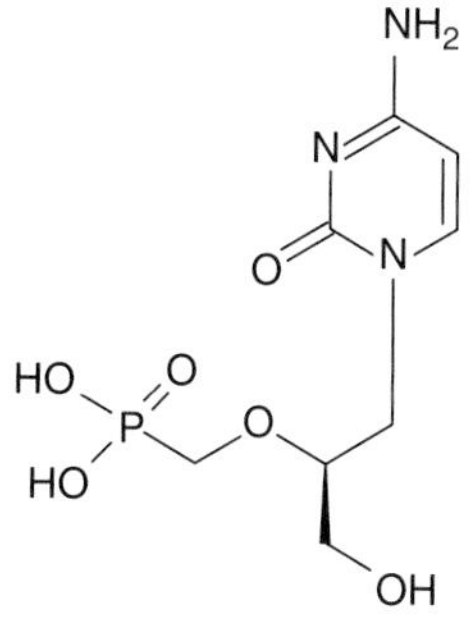

(*S*)-HPMPC
HPMPC
(*S*)-1-3-hydroxy-2-phosphonylmethoxypropylcytosine
Cidofovir
Vistide®

FIGURE 4

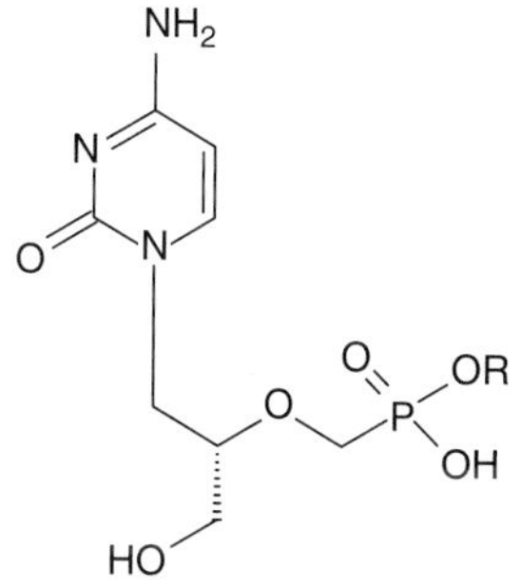

$R = O(CH_2)_3O(CH_2)_{15}CH_3$: HDP
$R = O(CH_2)_2O(CH_2)_{17}CH_3$: ODE

HDP-cidofovir
ODE-cidofovir

FIGURE 5

various experimental orthopox [i.e., cowpox, vaccinia, ectromelia (mousepox)] virus infections in mice (De Clercq, 2008a).

In the last few years, several new compounds have been described to inhibit orthopoxvirus replication at either a cellular target [ErbB-1 kinase: 4-anilinoquinazoline CI-1033 (Yang *et al.*, 2005a); Abl-family tyrosine kinase (Reeves *et al.*, 2005)] or a specific viral target [viral DNA polymerase: *N*-methanocarbathymidine (Smee *et al.*, 2007); virion assembly: mitoxantrone (Deng *et al.*, 2007)].

The most promising of these new compounds is ST-246 [4-trifluoromethyl-*N*-(3,3a,4,4a,5,5a,6,6a-octahydro-1,3-dioxo-4,6-ethanocycloprop[f] isoindol-2-(1*H*)-yl)benzamide (Fig. 6)]. ST-246 was shown in 2005 by Yang *et al.* (2005b) to be effective against multiple orthopoxviruses (i.e., vaccinia, monkeypox, camelpox, cowpox, ectromelia, and variola); ST-246 targets the F13L phospholipase involved in extracellular virus production and has demonstrated *in vivo* efficacy against systemic orthopoxvirus infection in mice (Quenelle *et al.*, 2007a; Sbrana *et al.*, 2007). ST-246 would confer synergistic efficacy in combination with HDP-cidofovir (Quenelle *et al.*, 2007b). In a recent case of severe eczema vaccinatum in a household contact (28-month-old son of a smallpox vaccinee), which finally resolved following successive treatment with intravenous vaccinia immune globulin, cidofovir, and ST-246, the latter may have helped contributing to the recovery process (Vora *et al.*, 2008).

IV. BENZIMIDAZOLE DERIVATIVES: SECOND ATTEMPT TO LAUNCH THE ANTIVIRAL CHEMOTHERAPY ERA

Following the thiosemicarbazones, the benzimidazoles could be considered as the second attempt to launch the era of antiviral chemotherapy (Tamm, 1956a). The Starting point was the inhibition of the multiplication

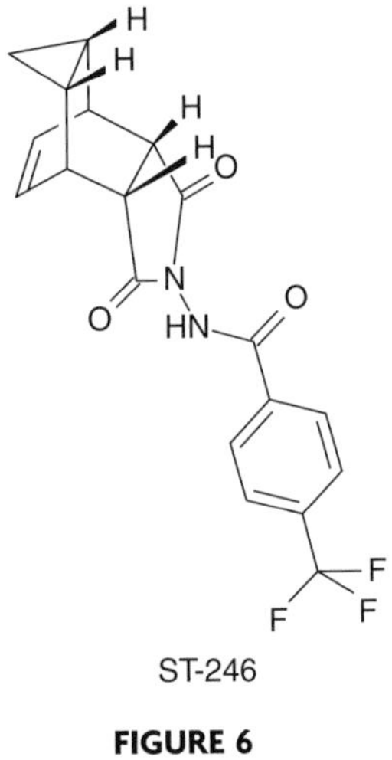

ST-246

FIGURE 6

of influenza A and B viruses in the chorioallantoic membrane of fertilized eggs by 2,5-dimethylbenzimidazole (Fig. 7). The reason for studying precisely this substance may seem bizarre: 2,5-dimethylbenzimidazole was apparently selected as it was considered an analog of a component of vitamin B12, namely 5,6-dimethylbenzimidazole, and vitamin B12 was closely involved in DNA synthesis and microbial growth. 2,5-Dimethyl-benzimidazole appeared to be unique in causing (bacterial cell) growth depression, and therefore, considered "suitable for beginning a study of the effects of benzimidazole derivatives on viral multiplication" (Tamm *et al.*, 1952).

Igor Tamm and his colleagues then realized that the 5,6-dimethylben-zimidazole moiety in vitamin B12 (as are the adenine and guanine moi-eties in nucleic acids) was linked to pentoses, which explains why the further tests were carried out, and inhibition of influenza virus multi-plication (in the chorioallantoic membrane) was found, with the 5,6-dichloro-1-β-D-ribofuranosylbenzimidazole (DRB) (Fig. 8) (Tamm and Tyrrell, 1954; Tamm *et al.*, 1954). DRB was quoted as causing inhibi-tion of influenza B (strain Lee) virus multiplication in intact embryonated

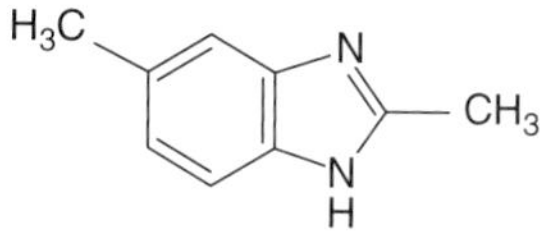

2,5-Dimethylbenzimidazole

FIGURE 7

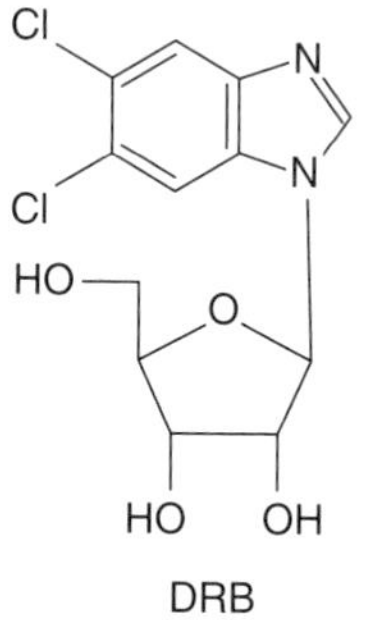

DRB
5,6-Dichloro-1-β-D-ribofuranosylbenzimidazole

FIGURE 8

chicken eggs and in mice without causing significant signs of toxicity in either host, chicken, or mice (Tamm *et al.*, 1954). The high inhibiting activity of DRB on influenza B virus multiplication was then extended to other halogenated, that is, 5- (or 6-)bromo-4,6- (or 5,7-dichloro-1-β-D-ribofuranosyl)benzimidazoles (Tamm *et al.*, 1956), the 4,5,6- (or 5,6,7-) trichloro-1-β-D-ribofuranosylbenzimidazole (TRB) being outstanding in both potency and selectivity (Tamm, 1956b).

The validity of the findings of Tamm *et al.* (1954a) that DRB inhibits the multiplication of influenza B virus in the chick embryo and in mice was questioned by Kissman *et al.* (1957). According to Horsfall Jr. and Tamm (1957), Kissman *et al.* (1957) even claimed that DRB ''has no effect on the multiplication of influenza virus.''

While Tamm *et al.* (1960) confirmed that DRB was effective against influenza virus multiplication, Tamm and his colleagues further showed that the inhibitory activity of benzimidazole derivatives extended to viruses other than influenza virus, such as vaccinia virus (Tamm and Overman, 1957), and poliovirus (Tamm and Nemes, 1957), although DRB and TRB were found to be less active against poliovirus than influenza virus.

A new benzimidazole derivative popping up in 1958 as being effective against polio(myelitis) virus was 2-(1-hydroxybenzyl)-benzimidazole or 2-(α-hydroxybenzyl)benzimidazole (HBB) (Fig. 9): it delayed or prevented poliomyelitis virus infection of mice (Hollingshead and Smith, 1958). Tamm and Nemes found that HBB was a highly active inhibitor of the multiplication of type 2 poliovirus in monkey kidney cells in culture (Tamm and Nemes, 1959), as reviewed by Igor Tamm at the occasion of a symposium on the experimental pharmacology and clinical use of anti-metabolites (Tamm, 1960). HBB is targeted at the replication of viral RNA (Eggers and Tamm, 1962). HBB is active against poliovirus, Coxsackie B virus, some Echo virus strains, and some Coxsackie A virus strains, but is devoid of activity against rhinoviruses, foot-and-mouth disease virus (FMDV), and hepatitis A virus (Eggers and Tamm, 1961). Initial mode of action studies on HBB (Eggers and Tamm, 1962) revealed that the

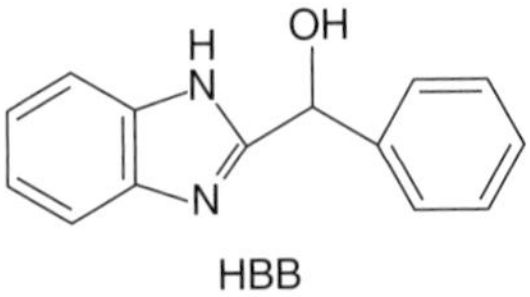

HBB
2-(α -Hydroxybenzyl)benzimidazole

FIGURE 9

compound does not interfere with early processes such as virus entry or uncoating. Eggers and Tamm (1963) then demonstrated that the production of poliovirus RNA was inhibited by HBB. Eggers (1976) demonstrated that HBB, when combined with guanidinium hydrochloride, was efficient in preventing virus-induced mortality in a lethal murine model of Echovirus type 9 or Coxsackie virus A9 infection.

With the successful implementation of the poliovirus vaccine, the search for selective inhibitors of picornavirus replication experienced a significant cutdown, but the continued occurrence of enterovirus infections, that is, Echo and Coxsackie virus infections, that could not be curtailed by vaccination, and the sheer fact that poliovirus, despite the most vigorous vaccination campaigns, has still not been eradicated, justifies the further search for selective inhibitors of picornavirus, and in particular, poliovirus replication (De Palma *et al.*, 2008).

V. RENAISSANCE OF THE BENZIMIDAZOLE DERIVATIVES: NOW TURNING INTO LEAD CANDIDATES FOR THE TREATMENT OF HUMAN CMV INFECTIONS

The "benzimidazole lead" was picked up by Karen Biron, John Drach, and Leroy Townsend, this time as a lead for specific inhibitors of human CMV infections. The prototype of these new benzimidazole derivatives, 1-(β-D-ribofuranosyl)-2-bromo-5,6-dichlorobenzimidazolep (BDCRB) (Fig. 10), despite its specific and potent activity against the human CMV, proved unsatisfactory for further development because of its rapid metabolism to the inactive and toxic aglycone (Biron *et al.*, 2002) (as reviewed by De Clercq, 2003).

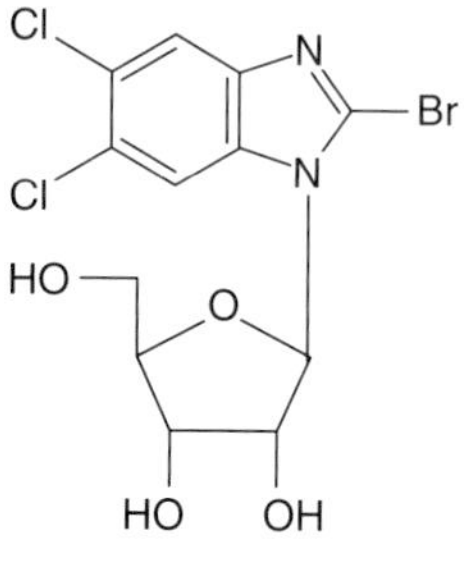

BDCRB
1-(β-D-ribofuranosyl)-2-bromo-5,6-dichlorobenzimidazole

FIGURE 10

Instead, the ʟ-counterpart of BDCRB, namely maribavir (1263W94), in which, in addition to the ᴅ → ʟ configuration switch, the bromine was replaced by an isopropylamine moiety (Fig. 11), was chosen for further development (De Clercq, 2003). Maribavir proved clearly more potent against CMV than either BDCRB or ganciclovir, and it was also less toxic than ganciclovir to bone marrow cells *in vitro* (Biron *et al.*, 2002). Curiously, maribavir appears to be targeted at the UL97 protein kinase, that is, the same enzyme that is required for the phosphorylation of ganciclovir to its monophosphate (Biron *et al.*, 2002). The UL97 kinase is an enzyme involved in the encapsidation and nuclear egress of CMV particles and the phosphorylation of virus replication-associated proteins. This means that in the anti-CMV effects of ganciclovir and maribavir, UL97 kinase plays opposite functions, that is, helping ganciclovir to get phosphorylated to its active metabolite so as to enable it to inhibit its target enzyme (i.e., the viral DNA polymerase), and directly acting as the target enzyme, as in the case of maribavir. CMV UL97 kinase mutations have been shown to confer resistance to maribavir (Chou *et al.*, 2007). These are different from the UL97 mutations linked to ganciclovir resistance. Resistance mutations in the UL97 gene conferring resistance to maribavir may arise swiftly (Chou and Marousek, 2008). Their clinical relevance remains to be determined (De Clercq, 2008a,c).

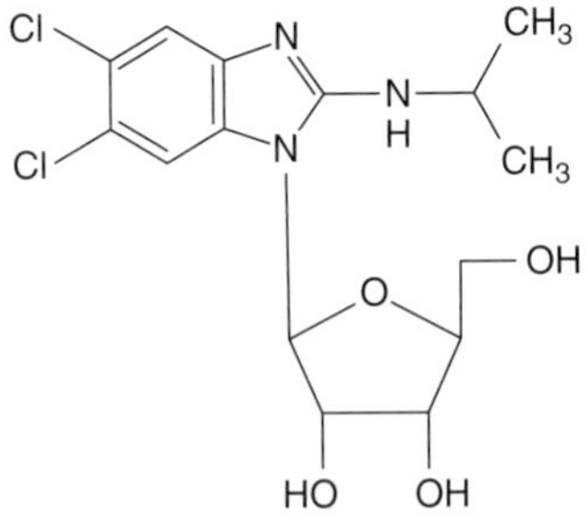

Maribavir
1263W94
1-(β-ʟ-ribofuranosyl)-2-isopropylamino-5,6-dichlorobenzimidazole

FIGURE 11

VI. 5-SUBSTITUTED 2'-DEOXYURIDINES: IDOXURIDINE (IDU) AND TRIFLURIDINE (TFT), THE THIRD AND DEFINITIVE ATTEMPT TO UNLEASH ANTIVIRAL CHEMOTHERAPY

While 5-iodouridine (IUR) had been synthesized first by Prusoff *et al.* (1953), the synthesis of its deoxyribosyl counterpart (5-iodo-2'-deoxyuridine, IUDR, IDU) (Fig. 12) was achieved only in 1959 when 2'-deoxyuridine became available in a sufficiently large quantity (Prusoff, 1959). At the onset, IDU was clearly considered as a potential antitumor agent, as evidenced by both the animal studies and clinical studies that were originally undertaken with 5-iodo-2'-deoxyuridine (Welch and Prusoff, 1960). Welch and Prusoff (1960) correctly pointed out that IUDR might be utilized in lieu of thymidine (5-methyl-2'-deoxyuridine) for the biosynthesis of DNA, the site of inhibition probably occurring at either the 5'-monophosphate or 5'-triphosphate level (Prusoff, 1960).

That IDU also had antiviral potential was first shown in 1961 by Ernest C. Herrmann Jr. who demonstrated that in an agar diffusion plaque-inhibition test, IDU inhibited plaque formation of DNA-containing viruses (vaccinia and HSV), but not RNA-containing viruses (West Nile and Newcastle disease virus); quite predictably, thymidine readily reversed the activity of IDU against vaccinia virus (Herrmann, 1961). The observations of Herrmann were quickly followed by a series of papers of Herbert Kaufman in 1962 reporting the prompt cure of well established lesions of HSV keratitis in rabbits upon topical application of

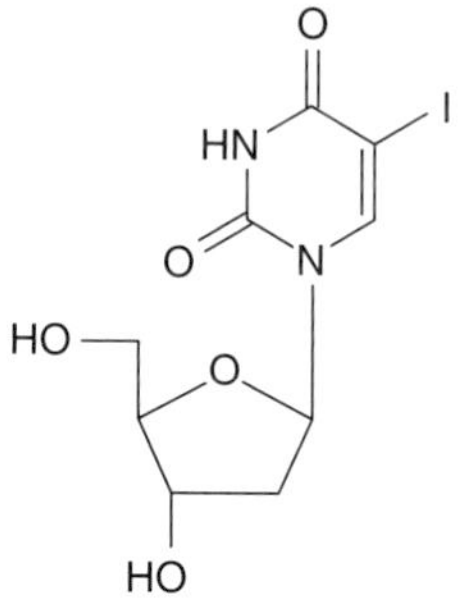

IDU
5-Iodo-2'-deoxyuridine
IUDR
Idoxuridine
Herpid®, Stoxil®

FIGURE 12

IDU, even when treatment was delayed for several days after infection (Kaufman, 1962; Kaufman *et al.*, 1962a). In the same year (1962) Kaufman and his colleagues reported the successful use of IDU in the (topical) treatment of HSV keratitis in humans (Kaufman *et al.*, 1962b). After Kaufman and his colleagues had documented the combined use of IDU and corticosteroids in the treatment of experimental herpetic keratitis in rabbits (Kaufman and Maloney, 1962), they proposed the combined use of IDU and corticosteroids in the treatment of the deeper manifestations (i.e., iritis) of herpetic eye infections in humans (Kaufman *et al.*, 1963).

In 1972, in their comprehensive review, Frank M. Schabel Jr. and John A. Montgomery referred to 22 favorable clinical reports on the effectiveness of topical IDU in the treatment of herpetic keratitis as opposed to only two negative reports, which meant that IDU was generally accepted by most observers to be unequivocally effective against herpetic keratitis, and so the prophecy of Kaufman *et al.* (1962b) that "IDU was the first specific chemotherapeutic agent effective against any "true" virus disease" was fulfilled (Schabel and Montgomery, 1972). In 1964, IDU would be joined by 5-trifluoromethyl-2′-deoxyuridine (trifluorothymidine, TFT) (Fig. 13) which was found effective in experimental herpetic keratitis caused by HSV strains that were either sensitive or resistant to IDU (Kaufman and Heidelberger, 1964). IDU and TFT would then become the first antivirals to be used, and are still in use, for the topical treatment of herpetic keratitis in humans (De Clercq, 2004a).

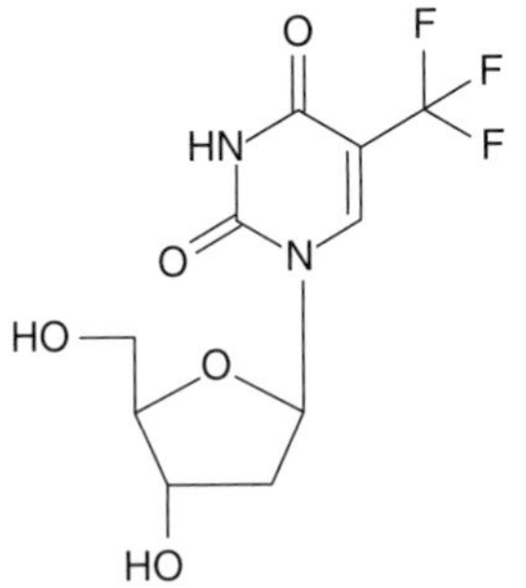

TFT

5-Trifluoromethyl-2′-deoxyuridine

Trifluorothymidine

Trifluridine

Viroptic®

FIGURE 13

While the effectiveness of IDU (and TFT) in the topical treatment (as eye drops or ointment) of herpetic keratitis could be considered as unequivocally established (Schabel and Montgomery, 1972), the effectiveness of IDU in treating cutaneous herpes, and most certainly, its effectiveness in the (systemic) treatment of herpes encephalitis, could, at best, be considered as tentative. Speculative anecdotal reports had indicated that systemic IDU might be effective in slowing or aborting HSV encephalitis (Breeden *et al.*, 1966), and Nolan *et al.* (1970, 1973) concluded that 4 out of 6, and later 9 out of 12 patients treated with idoxuridine for HSV encephalitis recovered from this devastating disease. However, two placebo-controlled studies with biopsy-proved cases of HSV encephalitis showed that at a dose of 100 mg/kg/day for 5 days, IDU failed to prevent death and gave unacceptable myelosuppression (Boston Interhospital Virus Study Group and the NIAID-Sponsored Cooperative Antiviral Clinical Study, 1975) and, consequently, the systemic use of IDU for the treatment of HSV encephalitis was no longer considered, and the potential use of TFT for such indication was not even envisaged.

VII. 5-SUBSTITUTED 2′-DEOXYURIDINES: IDU (AND TFT) AS THE STARTING POINT(S) FOR OTHER 5-SUBSTITUTED 2′-DEOXYURIDINES, THAT IS, BVDU [(*E*)-5-(2-BROMOVINYL)-2′-DEOXYURIDINE]

The initial successes achieved with IDU (and TFT) stimulated the interest in the synthesis of a myriad of new 5-substituted 2′-deoxyuridine derivatives (De Clercq, 1980), typical examples being those reported by Nemes and Hilleman (1965) (5-methylamino-2′-deoxyuridine)[81] and Kailash Gauri and David Shugar (5-ethyl-2′-deoxyuridine) (Gauri, 1968; Gauri and Malorny, 1967; Swierkowski and Shugar, 1969).

To see the advent of new clinically useful 5-substituted 2′-deoxyuridine, we had to wait until 1979 with the announcement of BVDU [brivudin, (*E*)-5-(2-bromovinyl)-2′-deoxyuridine (Fig. 14)] as a potent and selective inhibitor of HSV-1 (De Clercq *et al.*, 1979). As compared to IDU and TFT, BVDU appeared to be much "more gentle" in its approach to inhibit HSV-1 replication; it proved much more selective than IDU and TFT, in that its first phosphorylation step was carried out only by the virus-encoded thymidine kinase which imparted to BVDU a specificity only equaled by that of acyclovir (see *infra*). The unique advantage of BVDU, but also its weakness, was that its activity was restricted to HSV-1, and as it was apparent from the beginning, VZV as well (De Clercq *et al.*, 1980), but BVDU, unlike acyclovir (see *infra*), was virtually inactive against HSV-2. In the past (1980–1985), when strategic decisions had to be taken, lack of activity against HSV-2 was considered to be a serious

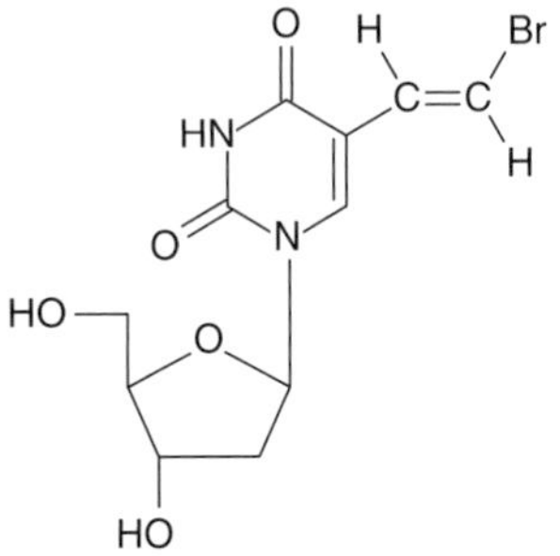

BVDU
(*E*)-5-(2-bromovinyl)-2'-deoxyuridine
Brivudin
Zostex®, Brivirac®, Zerpex®

FIGURE 14

handicap, and thus, while acyclovir swiftly moved toward it becoming the "gold standard" for the treatment of herpes (HSV-1 and HSV-2) infections, the development of BVDU lagged seriously behind to that of acyclovir, although from the beginning it was clear BVDU was far more superior than acyclovir in its activity against VZV. BVDU was finally marketed for the treatment of herpes zoster (orally at 125 mg once daily for 7 days), at least in several countries outside the United States and United Kingdom, but its primary indication, for which it was originally discovered, HSV-1 infections, should not be ignored, and this still makes BVDU a prime candidate for the treatment of herpetic HSV-1 keratitis, or herpes labialis, when applied topically, or mucocutaneous HSV-1 infections when applied systemically (i.e., orally at 125 mg once daily).

VIII. ARABINOSYLADENINE (ARA-A), ORIGINALLY CONCEIVED AS AN ANTITUMOR AGENT, THE FIRST ANTIVIRAL DRUG LICENSED AND USED FOR SYSTEMIC TREATMENT

The arabinosyl nucleoside analogs first mentioned in 1950 in extracts from sponges were arabinosyluracil (ara-U) and arabinosylthymine (ara-T) (Bergmann and Feeney, 1950). Their synthesis was described by Brown *et al.* (1956) and Fox *et al.* (1957). Then followed arabinosylcytosine (ara-C) and arabinosyladenine (ara-A). The first, ara-C, has been developed, and, as of today, is still used, as an anticancer agent. The second, ara-A (Fig. 15), first synthesized in 1960 by Lee *et al.* (1960), was originally considered to be a potential anticancer agent (Cohen, 1960; Schabel, 1968). Ara-A was first reported to be active, as an antiviral agent, against HSV and vaccinia virus by Privat de Garilhe and de Rudder (1964). Prior to the report of Privat de Garilhe and de Rudder (1964), Frank Schabel and his colleagues, including Robert ("Bob") W. Sidwell had independently observed the anti-HSV activity of ara-A in a microbial fermentation concentrate (references cited in Schabel, 1968).

Ara-A was the first of the nucleoside analogs considered to be sufficiently nontoxic to be given systemically, and Richard ("Rich") J. Whitley and his colleagues showed in 1976 that ara-A (vidarabine) was effective in the therapy of VZV infections (i.e., herpes zoster) in the immunosuppressed patients, providing that treatment commenced early in the disease (Whitley *et al.*, 1976). This was followed, one year later, by the demonstration that vidarabine was also effective in stopping the lethal progression of (biopsy-proved) herpes simplex encephalitis (Whitley *et al.*, 1977). After acyclovir had been launched in the 1980s for the treatment of HSV-1, HSV-2, and VZV infections, Whitley and his colleagues ran a controlled clinical trial comparing vidarabine with acyclovir in the treatment of neonatal HSV infection (Whitley *et al.*, 1991).

However, the major problem with vidarabine, besides its relative insolubility in aqueous medium, is its rapid deamination, by the ubiquitous adenosine deaminase to the antivirally inactive inosine counterpart, arabinosylhypoxanthine (ara-Hx). In attempts to prevent this deamination, various compounds were investigated as potential adenosine deaminase inhibitors, one of those compounds tried being acyclovir

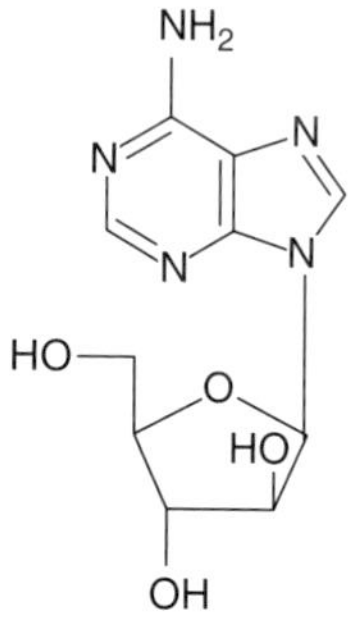

Arabinosyladenine
Ara-A
Vidarabine
Vira-A®

FIGURE 15

(originally called acycloguanosine), and when checked for its activity, as a control, acyclovir itself turned out to be antivirally active, as for the first time revealed by Peter Collins and John Bauer in the Wellcome Laboratories (Beckenham, UK) as early as 1974 according to a laboratory notebook (Field and De Clercq, 2004).

IX. ACYCLOVIR: THE START OF THE SELECTIVE ANTIVIRAL CHEMOTHERAPY ERA, AND STILL THE "GOLD STANDARD" FOR HSV THERAPY

The selectivity of antiviral action of acyclovir (Fig. 16), based on a specific recognition of acyclovir by the HSV-encoded thymidine kinase, was first announced in the December 1977 issue of the *Proceedings of the US National Academy of Sciences* (Elion *et al.*, 1977), the antiviral potential being further documented a few months later, in March 1978, in *Nature* (Schaeffer *et al.*, 1978). BVDU (first described in 1979 in *Proc. Natl. Acad. Sci. USA*) (De Clercq *et al.*, 1979) and acyclovir were the first truly selective antiviral agents, both reported at the end of the 1970s, and both exploiting the virus-encoded thymidine kinase to impart their selective mode of antiviral action. BVDU would eventually be marketed for the treatment of herpes zoster, while acyclovir would become the "gold standard" for the treatment of mucosal, cutaneous, and systemic HSV-1 and HSV-2 infections (including HSV encephalitis and genital herpes) as well as VZV infections (although acyclovir is much less potent (by 2 to 3 orders of magnitude) against VZV as compared to BVDU) (Field and De Clercq, 2008). Acyclovir is considered to be an extremely safe compound, which

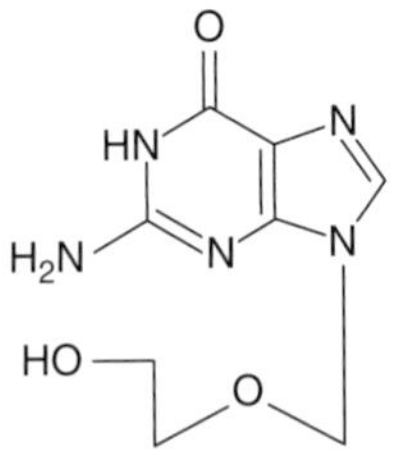

Acyclovir, aciclovir
9-(2-hydroxyethoxymethyl)guanine
Acycloguanosine
Zovirax®

FIGURE 16

may also be used for long-term suppressive therapy, for example in the prevention of recurrences of genital herpes.

Acyclovir has rather poor oral bioavailability. Therefore, for oral use, acyclovir has been replaced by its oral prodrug (valaciclovir) (Fig. 17), which is, in essence, used for the same indications as acyclovir (see above). Akin to the oral prodrug approach used for valaciclovir, a similar prodrug approach has been applied for ganciclovir (Fig. 18), in the treatment of CMV infections, and famciclovir (Fig. 19), in the treatment of HSV and VZV infections (De Clercq and Field, 2006).

There are, at present, in addition to acyclovir (and its prodrug, valaciclovir), only two other acyclic guanosine analogs which have been officially approved for the treatment of herpesvirus (HSV, VZV and/or CMV) infections. These are ganciclovir (and its prodrug valganciclovir) (Fig. 18) and penciclovir (and its prodrug famciclovir) (Fig. 19). Other herpesvirus infections such as Epstein–Barr virus (EBV), the causative agent of mononucleosis infectiosa, and human herpesvirus type HHV-6, which has been associated with several neurologic diseases, still await treatment with the appropriate antiviral agent, and *vice versa*, various acyclic guanosine analogs (such as H2G ((−)2HM-HBG), A-5021, cyclopropavir, and cyclohexenyl guanine) have been described (De Clercq *et al.*, 2001) that could be further developed as such, or in oral prodrug form (i.e., valomaciclovir stearate, the prodrug of H2G), for the treatment of a variety of herpesvirus infections (i.e., HSV-1, HSV-2, VZV, CMV, EBV, and HHV-6).

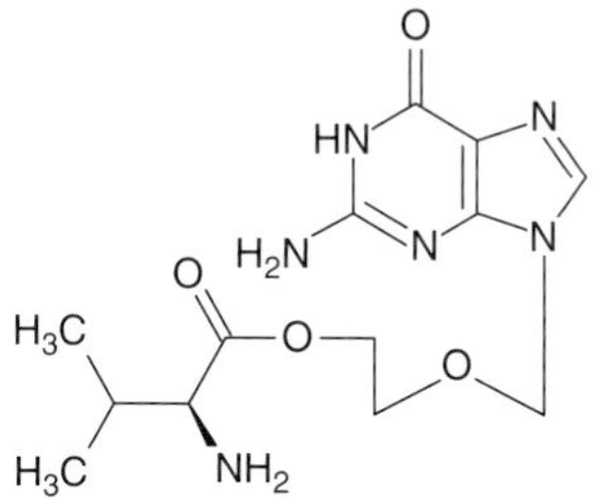

Valaciclovir
L-valine ester of 9-(2-hydroxyethoxymethyl)guanine
Valtrex®, Zelitrex®

FIGURE 17

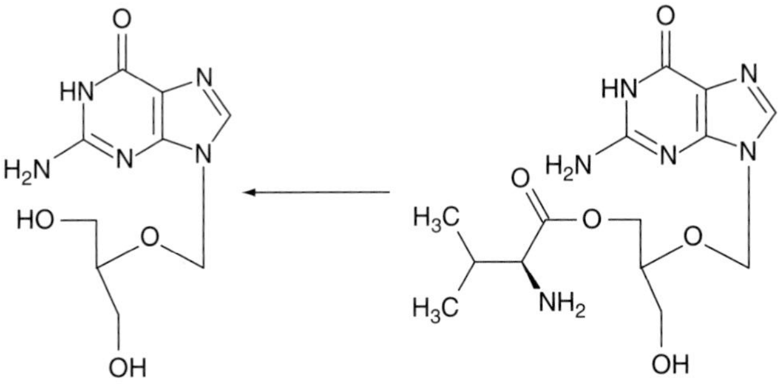

Ganciclovir
Cymevene®, Cytovene®

Valganciclovir
L-valine ester of 9-(1,3-dihydroxy-2-
propoxymethyl)guanine
Valcyte®

FIGURE 18

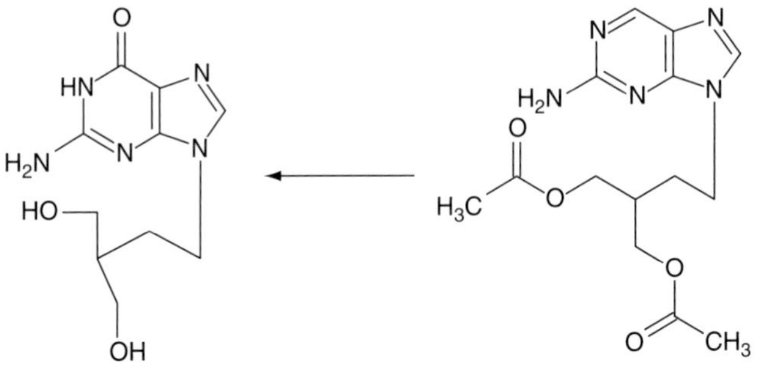

Penciclovir
Denavir®, Vectavir®

Famciclovir
Diacetyl 6-deoxy-9-(4-hydroxy-3-hydroxymethyl-
but-1-yl)guanine
Famvir®

FIGURE 19

X. ANTI-INFLUENZA VIRUS THERAPY: A FIRST ATTEMPT (DRB), FOLLOWED BY A SECOND (AMANTADINE) AND A THIRD ATTEMPT (NEURAMINIDASE INHIBITORS)

While earlier attempts (see *supra*) had pointed to the potential activity of DRB against influenza B virus, the first real anti-influenza virus agent ever to become an antiviral drug was amantadine (Fig. 20). It was first described in 1964, by Hoffmann and colleagues in *Science*, to be

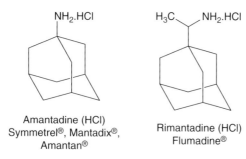

Amantadine (HCI)
Symmetrel®, Mantadix®,
Amantan®

Rimantadine (HCI)
Flumadine®

Adamantanamines

FIGURE 20

specifically effective, both *in vitro* and *in vivo* against influenza A (Davies *et al.*, 1964). It would later become clear, this was due to the fact that adamantanamine derivatives such as amantadine and rimantadine (Fig. 20) interacted with the M2 protein which is only part of influenza A (and does not occur in influenza B virus) (De Clercq, 2006a).

Although amantadine (and, later, rimantadine) became widely accepted for the treatment (and prophylaxis) of influenza A virus infections, and, for amantadine, the treatment of Parkinson's disease as well, this wide use of amantadine made the compound notorious for rapidly leading to virus–drug resistance (e.g., in avian influenza virus H5N1 strains) (He *et al.*, 2008). Numerous new adamantanamine derivatives have been synthesized by Nicolas Kolocouris and his colleagues of the University of Athens in Greece, as reviewed by De Clercq (2006a) and Lagoja and De Clercq (2008), but the problem still to be resolved is whether these new amantadine derivatives offer any advantage over amantadine in terms of *in vivo* efficacy, safety, and/or resistance profile.

Meanwhile, a new class of influenza (A and B) inhibitors emerged, thanks to the pioneering work of von Itzstein *et al.* (1993) and Kim *et al.* (1997), and these are the neuraminidase inhibitors zanamivir and oseltamivir (Fig. 21) (von Itzstein, 2007). These compounds block the release of newly formed influenza (A and B) virions from the cells, thus preventing their spread to other cells. Although new anti-influenza virus agents directed at new targets such as the viral RNA polymerase, may soon see the limelight, the neuraminidase inhibitors, and in particular, oseltamivir, better known as Tamiflu®, have been acclaimed as a new dimension in the treatment and/or prevention of influenza virus infections including the seasonal influenza A (H3N1 and H1N1) and influenza B infections as well as the much more feared avian influenza A H5N1 virus infections.

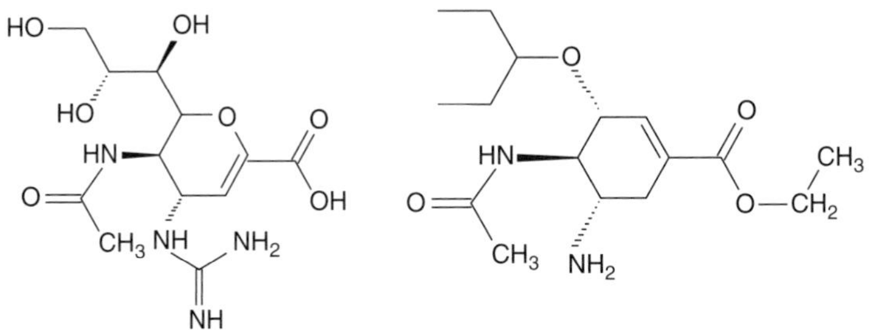

Zanamivir
Relenza®

Oseltamivir (ethyl ester)
Tamiflu®

FIGURE 21

The advantage of Tamiflu over Relenza is that Tamiflu can be administered orally by capsules, whereas Relenza has to be given by (oral) inhalation.

XI. RIBAVIRIN AND INTERFERON, TWO "OLD-TIMERS", JOINING FORCES IN THE TREATMENT OF A RELATIVELY NEW DISEASE, HEPATITIS C

Interferon, discovered by Isaacs and Lindenmann (1957) and ribavirin, first described by Sidwell et al. (1972) were at the time of their discovery in 1957 and 1972, respectively, both considered to be broad-spectrum antiviral agents. It could hardly be predicted at that time that so many years later, from the early 2000s, both compounds would be combined with success, in the treatment of a disease (hepatitis C) which was unknown at the time these compounds were discovered. In this combination used for the treatment of hepatitis C (Manns et al., 2007), interferon, mainly considered as an immunoregulator, primarily acts as an antiviral agent (De Clercq, 2004b), whereas ribavirin (Fig. 22), primarily viewed as an antiviral, may be principally acting as an immunosuppressive agent (Potter et al., 1976).

Standard therapy nowadays for the treatment of hepatitis C is based on the combination of pegylated interferon-α with ribavirin (Manns et al., 2007): about half of the patients are apparently cured [that is, achieving a sustained virological response (SVR)] following a 48-week course of pegylated interferon-α with ribavirin, although treatment is very hard, and patients often suffer from, along with extreme fatigue and flu-like symptoms, persistent headaches, muscle aches, and anemia. Whereas the flu-like syndrome can be attributed to interferon, anemia is obviously a side effect of ribavirin.

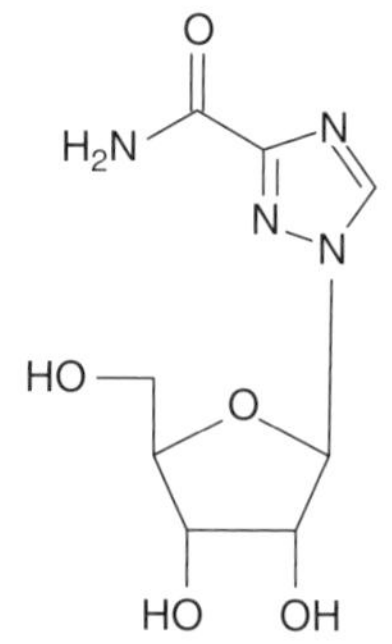

Ribavirin
1-β-ᴅ-ribofuranosyl-1H-1,2,4-triazole-3-carboxamide
Virazole®, Virazid®, Viramid®

FIGURE 22

Since 1985, ribavirin has been approved for the aerosol treatment of severe respiratory tract infections with respiratory syncytial virus (RSV) in children, although the actual clinical benefit achieved by ribavirin for this disease is still controversial. Another matter of debate is the action mechanism of ribavirin, although the most plausible explanation for both its antiviral and immunosuppressive action is, as was already pointed out by Streeter *et al.* (1973), an inhibitory effect on the IMP dehydrogenase, which converts IMP to XMP, with a concomitant suppression in the *de novo* biosynthesis of GMP, GDP, and GTP (Leyssen *et al.*, 2008).

XII. (S)-9-(2,3-DIHYDROXYPROPYL)ADENINE (DHPA), THE FIRST ACYCLIC ADENOSINE ANALOG, LEADING TO S-ADENOSYLHOMOCYSTEINE (SAH) HYDROLASE INHIBITORS AS BROAD-SPECTRUM ANTIVIRAL AGENTS

In 1978, within a few months after acyclovir had been described as a specific anti-HSV agent (Elion *et al.*, 1977; Schaeffer *et al.*, 1978), we reported in Science on the broad-spectrum antiviral activity of the acyclic adenosine analog, (S)-9-(2,3-dihydroxypropyl)adenine (DHPA) (Fig. 23) (De Clercq *et al.*, 1978). This compound originally synthesized by Anthonin Holý from the Institute of Organic Chemistry and Biochemistry (IOCB) in Prague, would later be marketed, albeit temporarily, in the former Czechoslovak Republic, under the trade name of Duviragel® for the topical treatment of cold sores (herpes labialis).

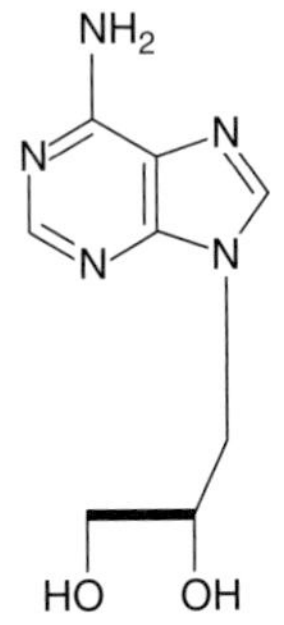

(S)-9-(2,3-dihydroxypropyl)adenine
(S)-DHPA
DHPA
Duviragel®

FIGURE 23

In 1983, DHPA was recognized as an inhibitor of S-adenosylhomocys-teine (SAH) hydrolase (Votruba *et al.*, 1983), and thus served as the prototype of a series of both acyclic and particularly carbocyclic adeno-sine analogs, that is, carbocyclic 3-deazaadenosine, neplanocin A, 3-dea-zaneplanocin A (Fig. 24), and their 5′-nor derivatives (De Clercq, 1985; De Clercq and Montgomery, 1983; De Clercq *et al.*, 1989); and their anti-viral activity, primarily encompassing poxviruses (i.e., vaccinia), (±)RNA viruses (i.e., reo) and (−)RNA viruses (i.e., bunya-, arena-, rhabdo-, filo-, orthomyxo-, and paramyxoviruses) correlated with, and could be attrib-uted to, their inhibitory effects on SAH hydrolase (De Clercq, 1987).

The SAH hydrolase inhibitors have not yet been used in the clinical setting. Yet, SAH hydrolase inhibitors have proved to be effective against Ebola virus in a lethal mouse model (Huggins *et al.*, 1999). Even when administered as a single dose of 1 mg/kg on the first or second day following an Ebola virus infection in mice, 3-deazaneplanocin A reduced peak viremia by more than 1,000-fold compared with mock-treated con-trols, and most or all of the animals survived (Bray *et al.*, 2000). This protective effect was accompanied, and probably be accounted for, by the massive production of interferon-α in the Ebola virus-infected mice treated with 3-deazaneplanocin A (Bray *et al.*, 2002). SAH hydrolase inhibitors such as 3-deazaneplanocin, block the 5′-capping of the nascent (+)RNA strands: this may prevent the dissociation of these (+)RNA strands from the viral (−)RNA templates, thus leading to the accumula-tion of replicative intermediates and as these replicative intermediates are

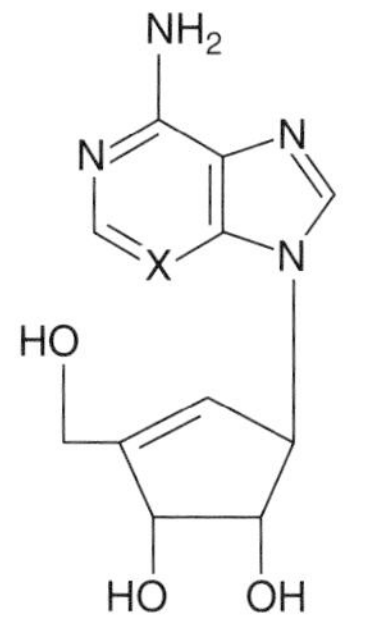

Neplanocin A (X = N)
3-deazaneplanocin A (X = CH)

FIGURE 24

partially composed of double-stranded RNA stretches, they may induce the high amounts of interferon-α under the given conditions (De Clercq, 2004b, 2008b).

XIII. (S)-9-(2,3-DIHYDROXYPROPYL)ADENINE (DHPA) LEADING TO THE FIRST ACYCLIC NUCLEOSIDE PHOSPHONATE, (S)-9-(3-HYDROXY-2-PHOSPHONYLMETHOXYPROPYL)ADENINE (HPMPA), AS A BROAD-SPECTRUM ANTIVIRAL AGENT

The first acyclic nucleoside phosphonate ever reported, preceding HPMPC (or cidofovir) (Fig. 4), was (S)-9-(3-hydroxy-2-phosphonyl-methoxypropyl)adenine (HPMPA) (Fig. 25). HPMPA could be conceived as a hybrid (De Clercq, 2008b) between phosphonoacetic acid (PAA) [from which the antiviral drug foscarnet (phosphonoformic acid, PFA) has been derived, which has proved useful upon intravenous injection in the treatment of acyclovir-resistant thymidine kinase deficient (TK⁻) HSV and VZV infections (Fig. 26)] and DHPA (see previous section). HPMPA, which, like DHPA, was synthesized by Anthonin Holý, was from its inception, considered to be a broad-spectrum anti-DNA virus agent with activity against all DNA viruses examined: herpes-, papilloma-, polyoma-, adeno-, pox-, and herpesviruses and among the herpesviruses all eight human herpesviruses: HSV-1, HSV-2, VZV, EBV, CMV, HHV-6, HHV-7, and HHV-8 (De Clercq *et al.*, 1986, 1987). Although HPMPA

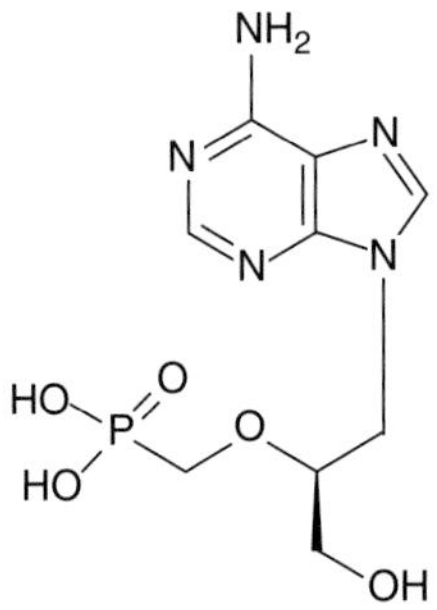

(S)-HPMPA
HPMPA
(S)-9-(3-hydroxy-2-phosphonylmethoxypropyl)adenine

FIGURE 25

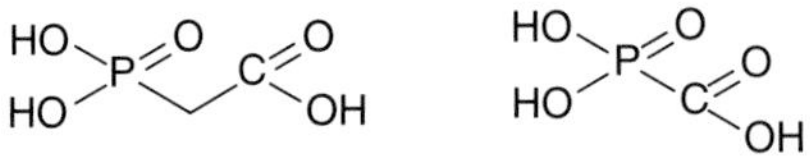

Phosphonoacetic acid   Phosphonoformic acid
PAA   PFA
                         Foscarnet
                         Foscavir®

FIGURE 26

(Fig. 25) was more potent than HPMPC (Fig. 4), the latter was developed as an antiviral drug because at a given stage it was considered as potentially less toxic.

XIV. 9-(2-PHOSPHONYLMETHOXYETHYL)ADENINE (PMEA), THE SISTER COMPOUND OF HPMPA

In 1986, in the same paper in *Nature* (De Clercq *et al.*, 1986) where we first reported on the broad spectrum anti-DNA virus activity of HPMPA, we also mentioned that a sister compound of HPMPA with a simpler, none-nantiomeric structure, namely 9-(2-phosphonylmethoxyethyl)adenine (PMEA, adefovir), later to be marketed in its oral prodrug form, bis (POM)PMEA (adefovir dipivoxil) (Fig. 27) had antiretrovirus activity and should be further pursued for its antiretrovirus potential.

Adefovir dipivoxil was initially pursued as an anti-HIV drug, and although it proved efficacious in the treatment of AIDS, as monitored by

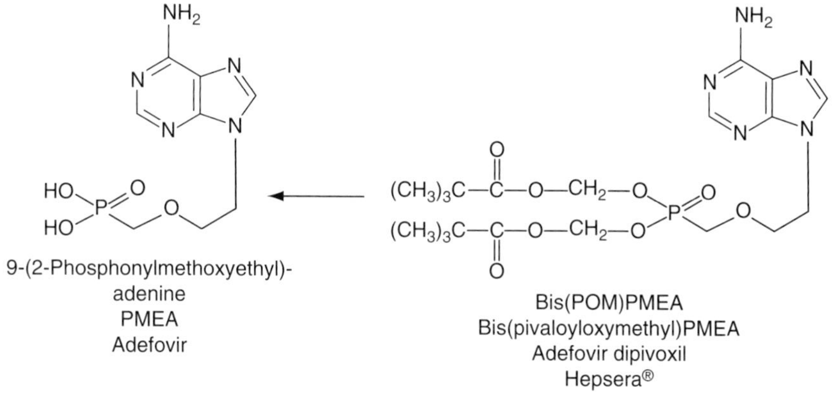

9-(2-Phosphonylmethoxyethyl)-
adenine
PMEA
Adefovir

Bis(POM)PMEA
Bis(pivaloyloxymethyl)PMEA
Adefovir dipivoxil
Hepsera®

FIGURE 27

a reduction in plasma viral load, it was considered too nephrotoxic to permit long-term use (>6 months) at the dosage (62.5 or 125 mg/day) required to inhibit HIV replication. Adefovir dipivoxil was then further pursued for the treatment of HBV infections, where it was demonstrated to be effective in reducing HBV DNA levels at a dosage (10 mg/kg/day) that was no longer toxic to the kidneys (or any other organs). In a number of papers in the *New England Journal of Medicine*, which have now become "classics," Hadziyannis *et al.* (2003) and Marcellin *et al.* (2003) clearly showed that adefovir dipivoxil was effective in the treatment of hepatitis B, whether HBV e antigen-positive or -negative, and a particularly nice crossing-over study further confirmed the efficacy of adefovir dipivoxil in the long-term therapy for HBeAg-negative chronic hepatitis B (Hadziyannis *et al.*, 2005).

Adefovir dipivoxil is effective against HBV infections that have developed resistance to lamivudine (3TC, for a long time was considered the drug of choice for the treatment of chronic hepatitis B) and, in the mean time, adefovir dipivoxil has become the drug of choice for the treatment of chronic hepatitis B. Resistance to adefovir may develop in HBV-infected patients, albeit at a much lower pace than resistance of HBV to lamivudine, that is, in 5.9% of the patients within 3 years (due to the reverse transcriptase mutations N236T or A181V) following adefovir (Hadziyannis *et al.*, 2005), as compared to >50% within 3 years for lamivudine.

In two double-blind phase 3 clinical studies conducted in patients with hepatitis B e antigen (HBeAg)-negative or HBeAg-positive chronic HBV infection over a period of 48 weeks, adefovir dipivoxil at a daily dose of

10 mg was superseded in antiviral efficacy by tenofovir disoproxil fumarate (TDF) at a daily dose of 300 mg (Marcellin *et al.*, 2008), which now makes TDF the drug of choice for the treatment of chronic hepatitis B.

XV. FROM PMEA (ADEFOVIR) TO PMPA (TENOFOVIR): IT ALL DEPENDS ON THE SUBSTITUTION OF A METHYL GROUP FOR A HYDROGEN

In 1993, we described for the first time (*R*)-9-(2-phosphonylmethoxypropyl)adenine (PMPA) (Fig. 28) and (*R*)-9-(2-phosphonylmethoxypropyl)-2,6-diaminopurine (PMPDAP) as antiretroviral agents (Balzarini *et al.*, 1993). Of these two compounds, PMPA was selected for further development, and as had been done for its predecessor PMEA, PMPA (tenofovir) was converted to an oral prodrug form, bis(POC)PMPA or tenofovir disoproxil (Naesens *et al.*, 1998; Robbins *et al.*, 1998). The latter was finally formulated as TDF and approved by the US Food and Drug Administration (FDA) for clinical use, for the treatment of HIV infections, in October 2001, within 8 years after it had been first described. Then followed the FDA approval for the combination of TDF with emtricitabine (Truvada®) in August 2004 and for the combination of TDF with emtricitabine and efavirenz (Atripla®) in July 2006.

The route from adefovir to Atripla®, via tenofovir, Viread®, and Truvada®, has been recently described (De Clercq, 2006b), as has been the importance of the phosphonate bridge in the antiviral activity of the acyclic nucleoside phosphonates (De Clercq, 2007; De Clercq and Holý, 2005). Of crucial importance in the final registration of Atripla® were the

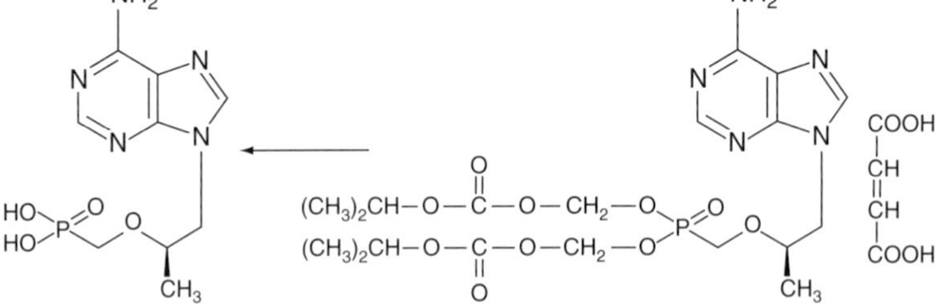

(*R*-PMPA
PMPA
(*R*)-9-(2-phosphonylmethoxy-
propyl)adenine
Tenofovir

Bis(POC)PMPA
Bis(isopropyloxycarbonyloxymethyl)PMPA
Tenofovir disoproxil fumarate
TDF
Viread®

FIGURE 28

studies of Gallant *et al.* (2006) and Pozniak *et al.* (2006) showing that over a 48- or 96-week treatment period, later extended to 144 weeks, the triple-drug combination of TDF with emtricitabine and efavirenz was superior to other triple-drug combinations, particularly the combination of zidovudine, lamivudine, and efavirenz, with regard to both efficacy (virologic and immunologic response) and safety (side effects).

What now urgently remains to be demonstrated is whether TDF taken orally as a single pill daily in either of its three forms (Viread®, Truvada®, or Atripla®) is also effective in the prophylaxis of HIV infection, irrespective of the route by which the virus is transmitted [sexually (i.e., via the vagina), parenterally (i.e., via needle stick), or perinatally (from mother to child)]. There is, first, ample evidence that experimentally tenofovir could prevent simian immunodeficiency virus (SIV) infection in macaques (Otten *et al.*, 2000; Tsai *et al.*, 1995; Van Rompay *et al.*, 2001); second, the prospects for an effective vaccine to prevent HIV infection seem to be remote as ever; and, third, a single oral pill daily (Viread®, Truvada®, or Atripla®) could be considered as a more convenient protective measure than, for example, a microbicidal gel, whose protection would be limited to its site of application.

XVI. SURAMIN, THE FIRST ANTIVIRAL DRUG EVER SHOWN TO INHIBIT HIV INFECTION BOTH *IN VITRO* AND *IN VIVO*

In 1979, I described suramin (Fig. 29) as a potent inhibitor of the reverse transcriptase of RNA tumor viruses (De Clercq, 1979). A few years later, in 1983, HIV [then called LAV for lymphadenopathy-associated virus or HTLV-III for human T-cell leukemia (or lymphotropic) virus type III] was identified as the tentative cause of AIDS, and suramin, because it was a reverse transcriptase inhibitor, was tested in 1984 and found active by Hiroaki Mitsuya and his colleagues against the *in vitro* infectivity of HTLV-III (Mitsuya *et al.*, 1984). By 1985, Sam Broder and his colleagues reported that suramin was also effective in suppressing virus (HTLV-III/LAV) levels in patients presenting with the AIDS-related complex (Broder *et al.*, 1985).

Suramin may have been further pursued as a potential drug for the treatment of AIDS [as it had already been used for the treatment of African trypanosimiasis (sleeping sickness) and onchocerciasis], but in that same year, 1985, azidothymidine became known as an inhibitor of HTLV-III/LAV infectivity, as demonstrated by Mitsuya *et al.* (1985). As azidothymidine (3′-azido-2′,3′-dideoxythymidine, AZT) from the beginning was perceived, rightfully, as being more potent and less toxic than suramin, and was soon to be followed by several other

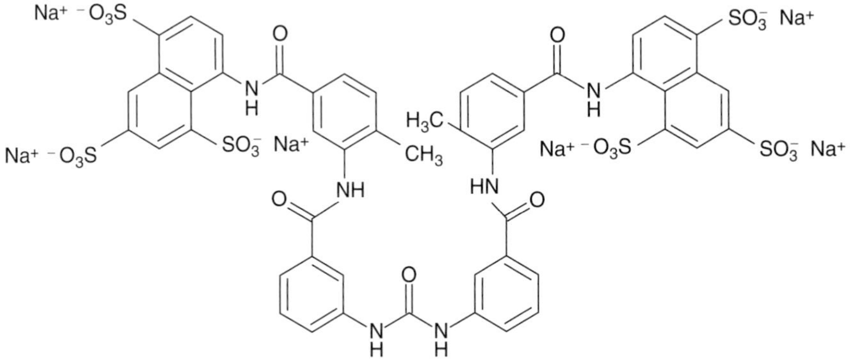

Suramin (hexasodium salt)
Moranyl, Naganol, Antrypol, Germanin, Bayer 205

FIGURE 29

2′,3′-dideoxynucleosides, such as 2′,3′-dideoxycytidine (ddC) and 2′,3′-dideoxyinosine as HTLV-III/LAV inhibitors (Mitsuya and Broder, 1986), suramin disappeared from the (anti-HIV drug) scene.

XVII. THE NUCLEOSIDE REVERSE TRANSCRIPTASE INHIBITORS (NRTIs) WITH AZIDOTHYMIDINE (AZT) AS THE STARTING POINT

Azidothymidine (AZT) was the first of the 2′,3′-dideoxynucleoside analogs (now commonly referred to as NRTIs or nucleoside reverse transcriptase inhibitors) to be recognized as potent and selective antiretroviral agents. The family of the NRTIs has now grown to seven compounds that have been officially licensed for systemic use in the treatment of HIV infections: AZT [azidothymidine, Retrovir®], ddC [zalcitabine (2′,3′-dideoxycytidine), Hivid®], ddI [didanosine (2′,3′-dideoxyinosine), Videx®], d4T [stavudine (2′,3′-dideoxy-2′,3′-didehydrothymidine), Zerit®], 3TC [lamivudine (3′-thia-2′,3′-dideoxycytidine), Epivir®], ABC (abacavir ((−)-(1S,4R)-4-[2-amino-6-(cyclopropylamino)-9H-purin-9-yl]-2-cyclo pentene-1-methanol), Ziagen®), and (−)FTC [emtricitabine ((−)-3′-thia-2′,3′-dideoxy-5-fluorocytidine), Emtriva®] (Fig. 30).

All the NRTIs act in a similar fashion in that they need to be phosphorylated to their active 5′-triphosphate form, before they are able to interact as competitive inhibitors/alternate substrates with the natural

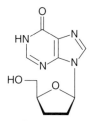

Zidovudine
3'-azido-2',3'-dideoxythymidine,
azidothymidine (AZT)
Retrovir®

Didanosine
2',3'-dideoxyinosine (ddI)
Videx®, Videx® EC

Zalcitabine
2',3'-dideoxycytidine (ddC)
Hivid®

Stavudine
2',3'-didehydro-2',3'-dideoxythymidine
(d4T)
Zerit®

Lamivudine
2',3'-dideoxy-3'-thiacytidine
3TC
Epivir®

Abacavir
(1S,4R)-4-[2-amino-6-(cyclopropylamino)-
9H-purin-9-yl]-2-cyclopentene-1-methanol
succinate (ABC)
Ziagen®

FIGURE 30 (Continued)

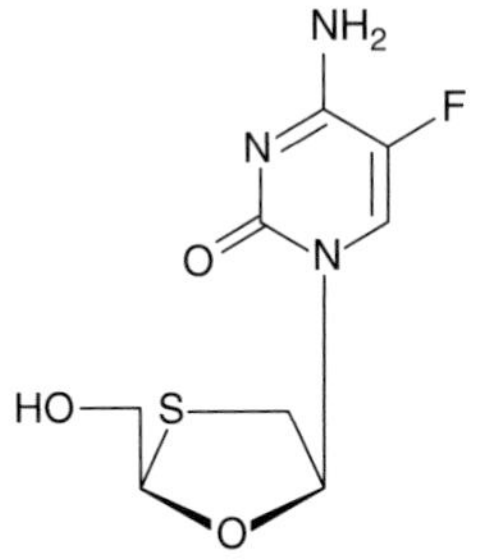

(−)-FTC
Emtricitabine
(−)-β-L-3′-thia-2′,3′-dideoxy-5-fluorocytidine
Emtriva®

FIGURE 30

substrates [dNTPs, dATP, dGTP, dCTP, or dTTP] in the reverse transcriptase (RT) reaction: for ddI, ddATP; for ABC, carbovir-TP; for ddC, 3TC, and (−)FTC, ddCTP, 3TC-TP, and (−)FTC-TP, respectively; and for AZT and d4T, AZT-TP and d4T-TP, respectively. As inhibitors, they prevent incorporation of the natural substrate into DNA, but as alternative substrates, they are themselves incorporated, and thus act as chain terminators, thereby preventing further chain elongation.

The NRTIs, with AZT (azidothymidine) as the prototype, should be clearly distinguished from the NtRTIs (i.e., nucleotide reverse transcriptase inhibitors) with tenofovir as the prototype, simply because when they are incorporated into the DNA chain, the NRTIs are incorporated as phosphates (i.e., ddTMPs, ddAMPs, ddCMPs, or ddGTPs), whereas the NtRTIs are incorporated as phosphonates (-P-C-O- instead of -P-O-C- for the phosphates), which makes their excision from the site of incorporation in the DNA much more difficult.

XVIII. THE NON-NUCLEOSIDE REVERSE TRANSCRIPTASE INHIBITORS (NNRTIs), WITH THE HEPT AND TIBO DERIVATIVES AS THE STARTING POINT

The era of the NNRTIs started in December 1989 with the description of the HEPT (Fig. 31) derivatives by Baba et al. (1989) and Miyasaka et al. (1989) as specific inhibitors of HIV-1. In 1991, it was ascertained (Baba et al., 1991a,b) that the HEPT {1-[(2-hydroxyethoxy)methyl]-6-(phenylthio)thymine} derivatives acted according to the current definition

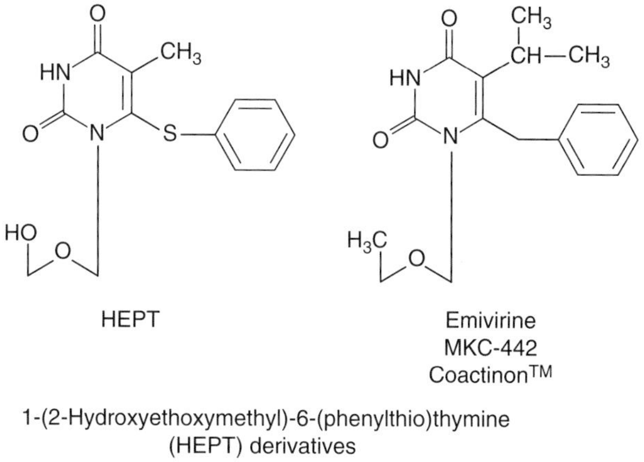

HEPT

Emivirine
MKC-442
Coactinon™

1-(2-Hydroxyethoxymethyl)-6-(phenylthio)thymine
(HEPT) derivatives

FIGURE 31

used for the class of the NNRTIs, that is, they specifically bind to a nonsubstrate binding (i.e., allosteric) site of the HIV-1 reverse transcriptase. The prototype of the HEPT derivatives, termed TS-II-25, was originally sent to our Laboratory in 1987 to be tested against HSV; and as it was found to be inactive against HSV, the story should have ended there, but then we found the compound to be active against HIV-1 (Baba *et al.*, 1989; Miyasaka *et al.*, 1989), and this ignited the whole area of the search for NNRTIs (De Clercq, 2008b). Further "lead" optimization studies led to the identification of MKC-442 (emivirine, Coactinon™) (Fig. 31) as a clinical candidate NNRTI (Baba *et al.*, 1994), and the compound progressed to advanced phase III clinical trials, before its further development was eventually abandoned.

At about the same time as the HEPT derivatives, we discovered the TIBO {tetrahydroimidazo[4,5,1-*jk*][1,4]-benzodiazepin-2(1*H*)-one and -thione}derivatives as potent and selective inhibitors of HIV-1 replication (Pauwels *et al.*, 1990). This discovery started from a collaborative effort initiated between Dr. Paul A.J. Janssen and our Laboratory in 1987. It was based upon the rational screening of about 600 compounds from the Janssen Library, complemented by lead optimization through chemical modifications, and led to identification of TIBO R82150 and later 8-chloro-TIBO R86183 (tivirapine) (Fig. 32) as the prototype compounds. In their mode of action the TIBO derivatives behaved very much like the HEPT derivatives, and with some imagination, the TIBOs (i.e., tivirapine) and HEPTs (i.e., emivirine) could be considered as sharing overlapping features (De Clercq, 2004c).

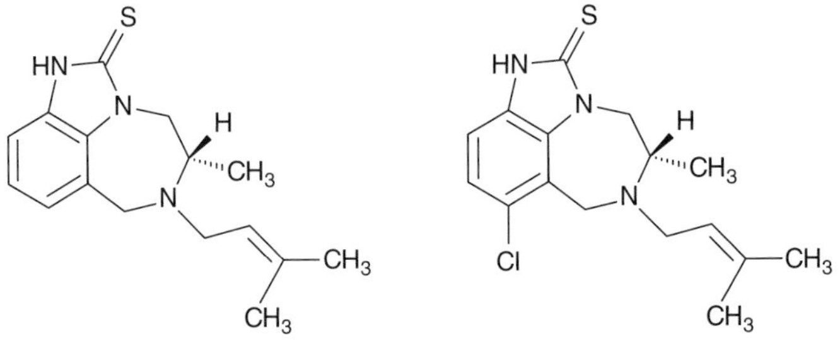

TIBO R82150   8-Chloro-TIBO R86183

Tetrahydroimidazo[4,5,1-jk][1,4]-benzodiazepin-2(1H)-one
(TIBO) derivatives

FIGURE 32

Although the original HEPT and TIBO derivatives, that is, emivirine and tivirapine, respectively, were eventually not commercialized for clinical use, they paved the way for other NNRTIs to be effectively marketed as anti-HIV-1 drugs, in particular nevirapine, delavirdine, efavirenz, and etravirine (Fig. 33), which, in their mode of action, followed the same pattern as proposed for the original NNRTIs, HEPT, and TIBO; that is, they bind to an allosteric (nonsubstrate) binding site of the HIV-1 reverse transcriptase, albeit with a greater resilience for the NNRTI signature resistance mutations (K103N and Y181C) for the "newer" NNRTIs (i.e., etravirine) than for the "older" NNRTIs (i.e., nevirapine).

Although not (yet) approved for clinical use, the "newest" among the NNRTIs, rilpivirine (Fig. 34), first revealed by Janssen *et al.* (2005) fulfills virtually all requirements for a successful anti-HIV drug: ease of synthesis and formulation, high potency even against HIV-1 mutants (i.e., K103N and Y181C) resistant to other NNRTIs, oral bioavailability, and prolonged duration of activity. Rilpivirine may eventually come close to fulfilling Dr. Paul's ultimate dream to develop the "miracle" drug for the treatment of AIDS.

XIX. THE HIV PROTEASE INHIBITORS (PIs), HAILED FROM THEIR INCEPTION, AS RESULTING FROM RATIONAL DESIGN

There are, at present, 10 anti-HIV compounds (Fig. 35) officially licensed for the treatment of HIV infections: saquinavir, ritonavir, indinavir, nelfinavir, amprenavir, lopinavir, atazanavir, fosamprenavir, tipranavir, and

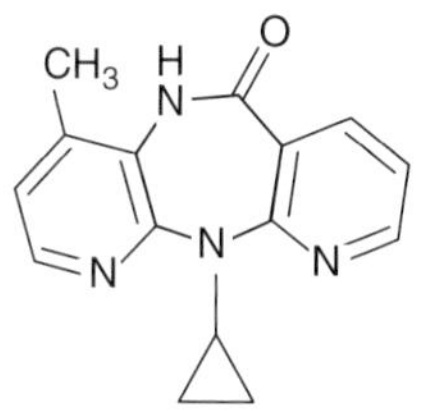

Nevirapine
Viramune®

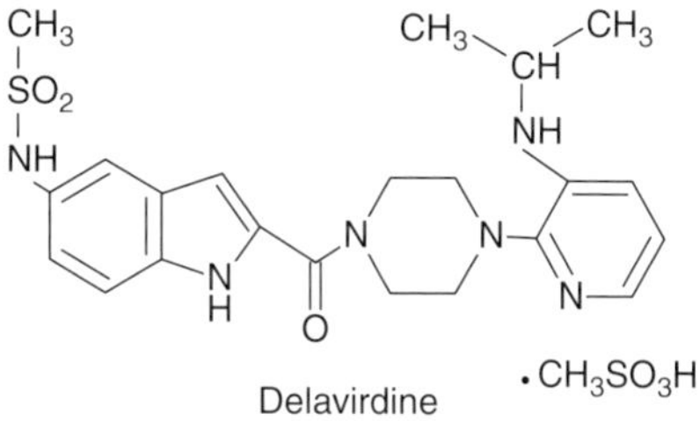

Delavirdine
Rescriptor®

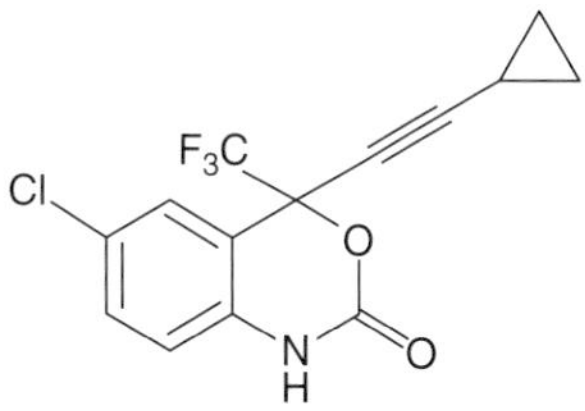

Efavirenz
Sustiva®, Stocrin®

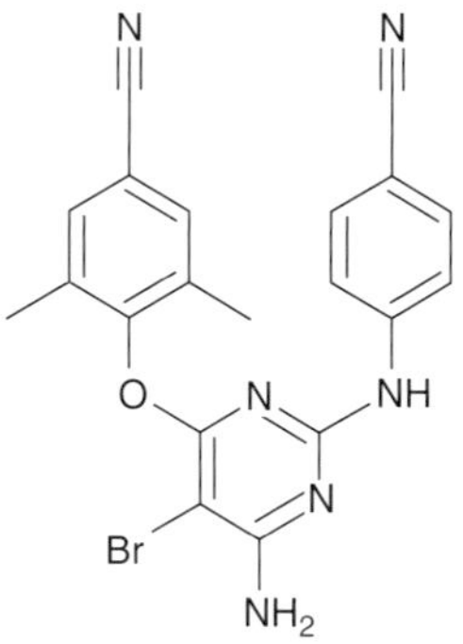

Etravirine (TMC125, R165335)
Intelence®

FIGURE 33

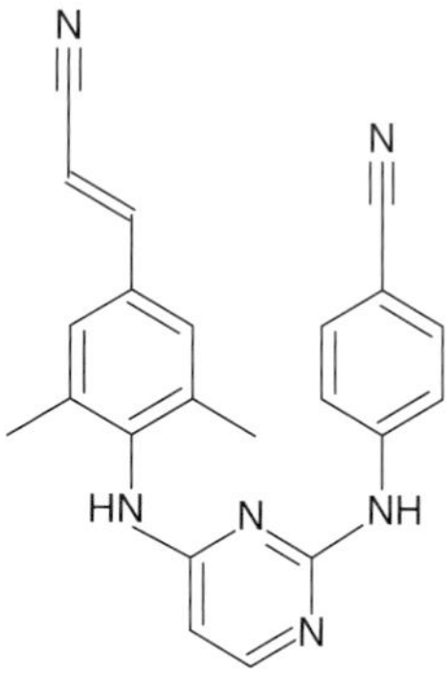

Rilpivirine
(TMC278, R278474)

FIGURE 34

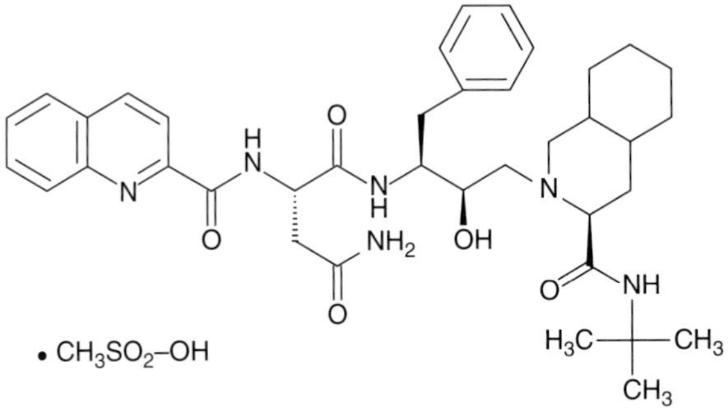

• CH₃SO₂–OH

Saquinavir
hard gel capsules, Invirase®
soft gelatin capsules, Fortovase®

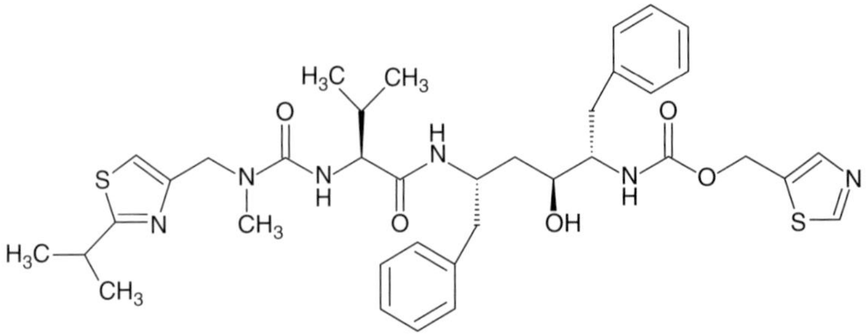

Ritonavir
Norvir®

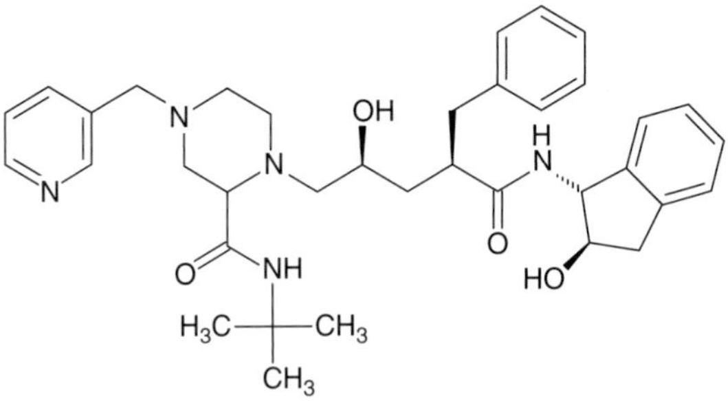

Indinavir
Crixivan®

FIGURE 35   (Continued)

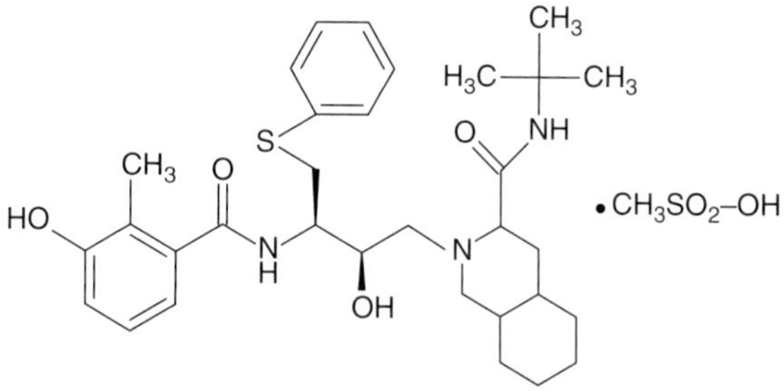

Nelfinavir
Viracept®

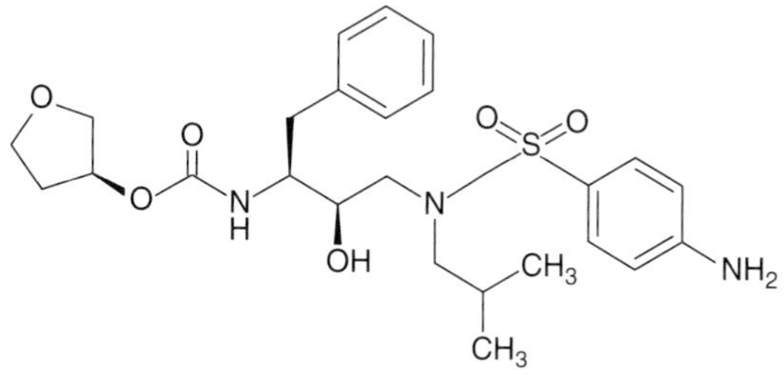

Amprenavir
Agenerase®, Prozei®

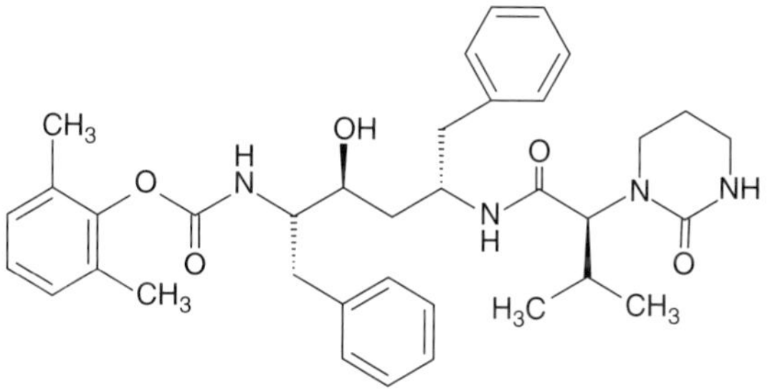

Lopinavir
combined with ritonavir at 4/1 ratio
Kaletra®

FIGURE 35   (Continued)

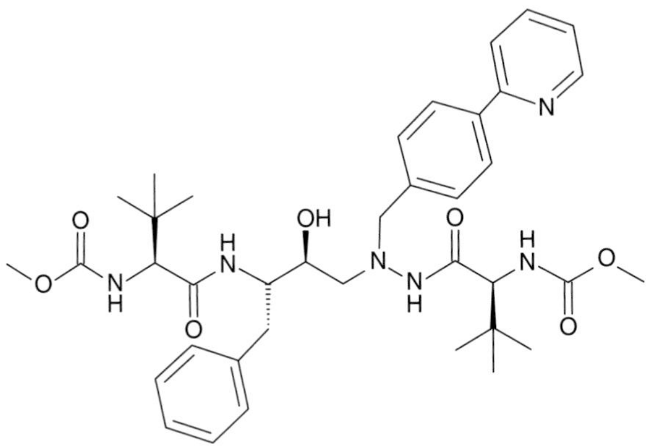

Atazanavir
Reyataz®

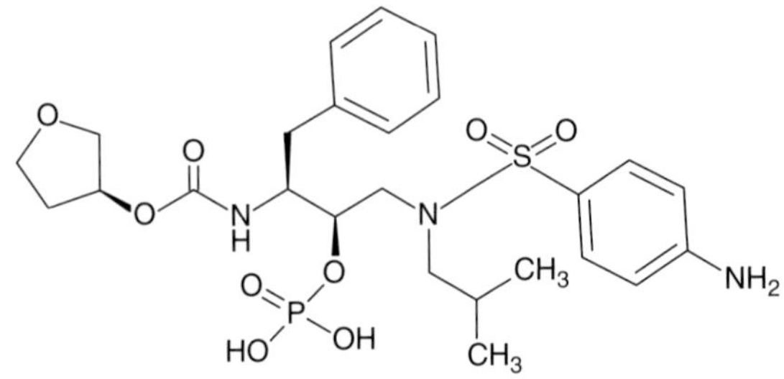

Fosamprenavir
Lexiva®, Telzir®

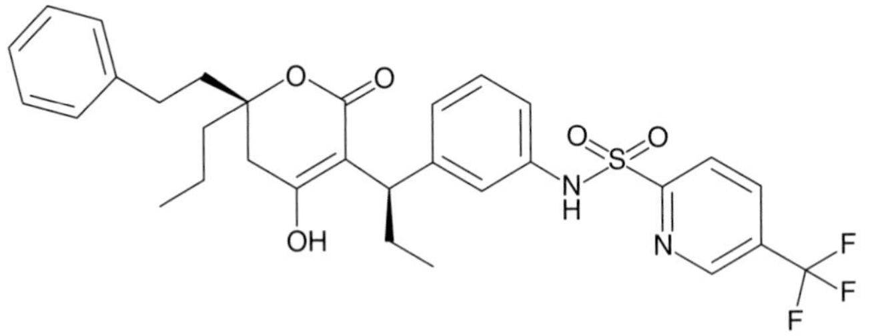

Tipranavir (U-140690)
Aptivus®

FIGURE 35   (Continued)

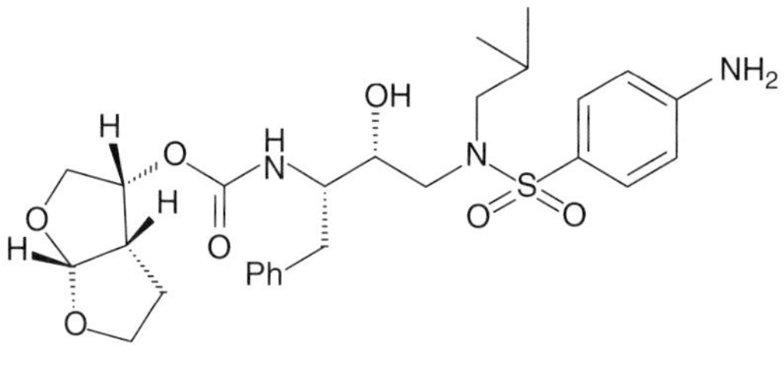

Darunavir (TMC-114)
Prezista®

FIGURE 35

darunavir (Fig. 35). Except for tipranavir, which is a coumarin derivative, all other compounds can be considered as peptidomimetics, in that, instead of the normal [–NH–CO–] peptide linkage, they contain the peptidomimetic hydroxyethylene $\left[-CH_2-\overset{\displaystyle OH}{\underset{\displaystyle |}{CH}}- \right]$ bond, which cannot be cleaved by the HIV protease, and, thereby "fooling" and inactivating the viral protease.

In contrast with the NRTIs and NNRTIs which were also discovered from so-called "screening" procedures using cell-based assays, the protease inhibitors (PIs) were claimed as rationally designed (Roberts *et al.*, 1990). In fact, the HIV protease being an aspartyl protease, much experience was gained from the insight into the molecular mode of action of other aspartyl protease (i.e., renin) inhibitors (Greenlee, 1990), and recombinant DNA technology offered the opportunity to clone, express, and purify the HIV protease (Graves *et al.*, 1988; Mous *et al.*, 1988), thus allowing the initial testing of potential inhibitors. This not only led to the discovery of saquinavir, the first of the PIs (Roberts *et al.*, 1990), but also of ritonavir, indinavir, and all other peptidomimetic HIV protease inhibitors (Dorsey and Vacca, 2001; Duncan and Redshaw, 2001; Erickson, 2001; Kempf, 2001).

The 10 PIs which in the meantime have been licensed for clinical use (Fig. 35) share the same mode of action, in that they prevent the cleavage of the precursor *gag* and *gag–pol* proteins into the mature *gag* (capsid) and *pol* (protease, reverse transcriptase, and integrase proteins): this process is initiated by the protease, which therefore has to be cleaved autocatalytically from the *gag–pol* precursor protein. If this proteolytic cleavage is blocked, that is, by a given PI, infectious HIV particles are not produced, and virus spread is stopped. PIs have been shown to fit snugly within the active site of the (dimeric) HIV protease (Pauwels, 2006).

XX. NEW HIV INHIBITORS, TARGETED AT EITHER FUSION (ENFUVIRTIDE), CORECEPTOR USAGE (MARAVIROC), OR INTEGRASE (RALTEGRAVIR)

The search for new targets that could be exploited successfully in the design and development of new anti-HIV drugs has revealed (i) the HIV–cell fusion process, (ii) the interaction of HIV with its coreceptor (CCR5), and (iii) the HIV integrase as appropriate sites for chemotherapeutic attack. This search only took off seriously in the 1990s after the virus–cell fusion process became better known, the role of HIV integrase better defined, and the coreceptors for HIV had been identified (which occurred precisely in 1996). There is one fusion inhibitor (FI) currently available for the treatment of HIV infections, namely enfuvirtide (Fig. 36), which corresponds to a polypeptide of 36 amino acids that is homologous to, and engages in a coil–coil interaction with, the heptad repeat (HR) regions of the viral envelope glycoprotein gp41 (Matthews *et al.*, 2004). Consequently, the fusion of the virus particle with the outer cell membrane is blocked. Enfuvirtide is the only anti-HIV compound that has a polypeptidic structure, and, hence, is not orally bioavailable: it must be injected parenterally (subcutaneously, twice daily).

Coreceptor inhibitors (CRIs) antagonize the interaction of the coreceptors (CCR5 or CXCR4) used by, respectively, M (macrophage-tropic) and T (lymphocyte-tropic) HIV strains (now generally termed R5 and X4 strains, respectively) to enter their target cells. To enter these cells, HIV through its envelope glycoprotein gp120 first binds to its primary receptor (CD4) before it interacts, again through its glycoprotein gp120, with the coreceptor (CCR5 or CXCR4). Several CCR5 and CXCR4 inhibitors have been described during the last 10 years, the only coreceptor inhibitor that has been licensed for clinical use being the CCR5 antagonist maraviroc (Fig. 37) (Perros, 2007). The major problem with maraviroc and other potentially forthcoming CCR5 antagonists is that they are active only against R5 HIV strains, and thus may stimulate the selection of X4 strains from a mixed R5/X4 population. Ideally, a CCR5 antagonist should be combined with a CXCR4 antagonist so as to block both coreceptors at the same time (De Clercq, 2009a).

Although integrase has been considered an attractive target for potential anti-HIV drugs for circa 15 years (Pommier *et al.*, 2005), thanks to the pioneering work of Hazuda *et al.* (2000, 2004), an adequate clinical candidate integrase inhibitor, MK-0518 (raltegravir) (Fig. 38) was brought forward, and eventually licensed, in October 2007, for the treatment of HIV-1 infection (Cooper *et al.*, 2008; Grinsztejn *et al.*, 2007; Steigbigel *et al.*, 2008). As is "de rigeur" for all other anti-HIV drugs, raltegravir (and other forthcoming integrase inhibitors) will have to be used in combination

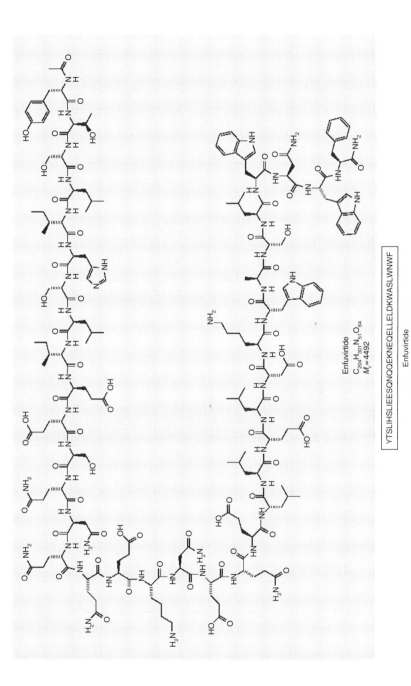

YTSLIHSLIEESQNQQEKNEQELLELDKWASLWNWF

Enfuvirtide
$C_{204}H_{301}N_{51}O_{64}$
$M_r = 4492$

Enfuvirtide
DP-178, pentafuside, T20
Fuzeon®

FIGURE 36

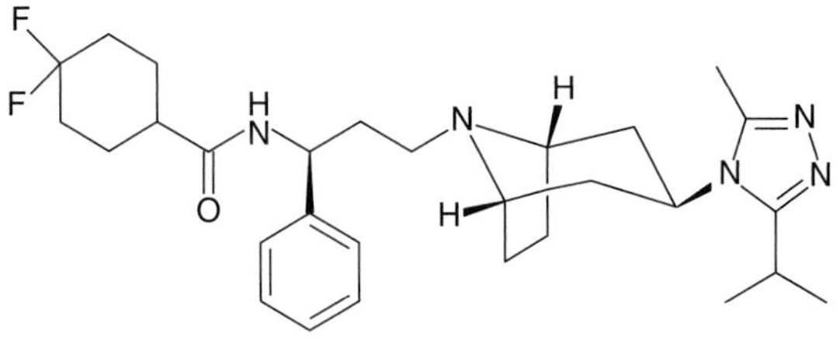

UK-427857
Maraviroc
Selzentry®

FIGURE 37

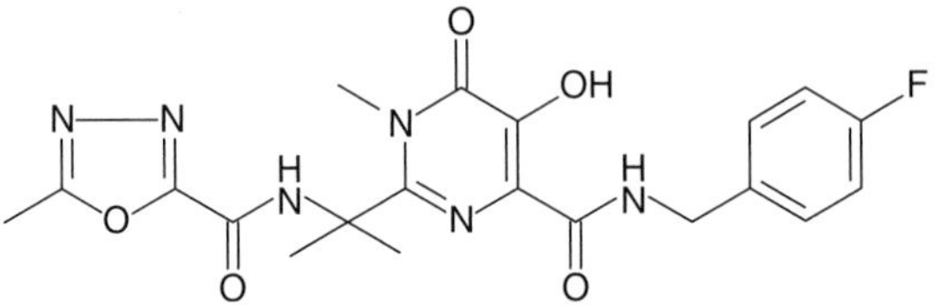

MK-0518
Raltegravir
Isentress®

FIGURE 38

with other anti-HIV drugs targeted at other enzymes (i.e., reverse transcriptase or HIV protease) so as to minimize the risk of resistance development (De Clercq, 2009b).

XXI. CONCLUSION

If we count all the antiviral drugs that have been formally licensed (although not necessarily still used) for the treatment of virus infections: 25 compounds[175] have been approved, within 25 years after the discovery of HIV (then called HTLV-III/LAV), for the treatment of HIV infections; the other 25 compounds have been formally approved for the treatment of HSV, VZV, CMV, HBV, HCV, or influenza virus infections. The compounds that have been formally approved for the treatment of HSV infections are acyclovir, valaciclovir, famciclovir, penciclovir, idoxuridine, trifluridine (the latter three only for topical application), and vidarabine (no longer used); for the treatment of VZV infections, acyclovir, valaciclovir, famciclovir, and brivudin; for the treatment of CMV infections, ganciclovir, valganciclovir, foscarnet, cidofovir, and fomivirsen (the latter only for intravitreal injection); for the treatment of chronic HBV infections, lamivudine, adefovir dipivoxil, entecavir, telbivudine, clevudine, and tenofovir disoproxil fumarate (TDF); for the treatment of chronic HCV infections, ribavirin in combination with (pegylated) interferon-α; and for the treatment of influenza virus infections, amantadine, rimantadine, oseltamivir, and zanamivir (Field and De Clercq, 2008; De Clercq and Field, 2008a,b).

For the treatment of HIV infections, exactly 25 anti-HIV compounds (De Clercq, 2009b) were formally approved by the end of 2008. These compounds fall into six categories: nucleoside reverse transcriptase inhibitors (NRTIs: zidovudine, didanosine, zalcitabine, stavudine, lamivudine, abacavir, and emtricitabine), nucleotide reverse transcriptase inhibitors (NtRTIs: tenofovir in its oral prodrug form; tenofovir disoproxil fumarate (TDF)); non-nucleoside reverse transcriptase inhibitors (NNRTIs: nevirapine, delavirdine, efavirenz, and etravirine); protease inhibitors (PIs: saquinavir, ritonavir, indinavir, nelfinavir, amprenavir, lopinavir, atazanavir, fosamprenavir, tipranavir, and darunavir); cell entry inhibitors [fusion inhibitors (FIs: enfuvirtide) and coreceptor inhibitors (CRIs: maraviroc)]; and integrase inhibitors (INIs: raltegravir).

There are, in addition to the 50 antiviral compounds that have been presently licensed for clinical use, numerous others emerging as antiviral drugs for the treatment of the virus infections mentioned above as well as other virus infections (De Clercq, 2008a,c). It is likely that some of these emerging antivirals may become real drugs in the future, and that the antiviral drug era may further expand in the next few years, much like the antibiotics did in the past with, hopefully, less detrimental consequences in terms of drug resistance development.

ACKNOWLEDGMENT

I am most indebted to Christiane Callebaut for her proficient editorial assistance.

REFERENCES

Ackermann, W. W. (1952). Aminosulfonic acids and viral propagation. *Proc. Soc. Exp. Biol. Med.* **80**:362–367.
Ackermann, W. W., and Maassab, H. F. (1954). Growth characteristics of influenza virus: The influence of a sulfonic acid. *J. Exp. Med.* **99**:105–117.
Baba, M., Tanaka, H., De Clercq, E., Pauwels, R., Balzarini, J., Schols, D., Nakashima, H., Perno, C.-F., Walker, R. T., and Miyasaka, T. (1989). Highly specific inhibition of human immunodeficiency virus type 1 by a novel 6-substituted acyclouridine derivative. *Biochem. Biophys. Res. Commun.* **165**:1375–1381.

Baba, M., De Clercq, E., Tanaka, H., Ubasawa, M., Takashima, H., Sekiya, K., Nitta, I., Umezu, K., Nakashima, H., Mori, S., Shigeta, S., Walker, R. T., *et al.* (1991a). Potent and selective inhibition of human immunodeficiency virus type 1 (HIV-1) by 5-ethyl-6-phenylthiouracil derivatives through their interaction with the HIV-1 reverse transcriptase. *Proc. Natl. Acad. Sci. USA* **88:**2356–2360.

Baba, M., De Clercq, E., Tanaka, H., Ubasawa, M., Takashima, H., Sekiya, K., Nitta, I., Umezu, K., Walker, R. T., Mori, S., Ito, M., Shigeta, S., *et al.* (1991b). Highly potent and selective inhibition of human immunodeficiency virus type 1 by a novel series of 6-substituted acyclouridine derivatives. *Mol. Pharmacol.* **39:**805–810.

Baba, M., Shigeta, S., Yuasa, S., Takashima, H., Sekiya, K., Ubasawa, M., Tanaka, H., Miyasaka, T., Walker, R. T., and De Clercq, E. (1994). Preclinical evaluation of MKC-442, a highly potent and specific inhibitor of human immunodeficiency virus type 1 *in vitro*. *Antimicrob. Agents Chemother.* **38:**688–692.

Balzarini, J., Holý, A., Jindrich, J., Naesens, L., Snoeck, R., Schols, D., and De Clercq, E. (1993). Differential antiherpesvirus and antiretrovirus effects of the (*S*) and (*R*) enantiomers of acyclic nucleoside phosphonates: Potent and selective *in vitro* and *in vivo* antiretrovirus activities of (*R*)-9-(2-phosphonomethoxypropyl)-2, 6-diaminopurine. *Antimicrob. Agents Chemother.* **37:**332–338.

Bauer, D. J. (1955). The antiviral and synergistic actions of isatin thiosemicarbazone and certain phenoxypyrimidines in vaccinia infection in mice. *Br. J. Exp. Pathol.* **36:**105–114.

Bauer, D. J. (1972). Thiosemicarbazones. *In* "Chemotherapy of Virus Diseases" (D. J. Bauer, ed.), Vol. 1, pp. 35–113. Pergamon Press, Oxford.

Bauer, D. J., and Sadler, P. W. (1960). The structure-activity relationships of the antiviral chemotherapeutic activity of isatin β-thiosemicarbazone. *Br. J. Pharmacol. Chemother.* **15:**101–110.

Bauer, D. J., Kempe, C. H., and Downie, A. W. (1963). Prophylactic treatment of smallpox contacts with N-methylisatin β-thiosemicarbazone. *Lancet* **ii:**494–496.

Bergmann, W., and Feeney, R. J. (1950). The isolation of a new thymine pentoside from sponges. *J. Am. Chem. Soc.* **72:**2809–2810.

Biron, K. K., Harvey, R. J., Chamberlain, S. C., Good, S. S., Smith, A. A., 3rd, Davis, M. G., Talarico, C. L., Miller, W. H., Ferris, R., Dornsife, R. E., Stanat, S. C., Drach, J. C., *et al.* (2002). Potent and selective inhibition of human cytomegalovirus replication by 1263W94, a benzimidazole L-riboside with a unique mode of action. *Antimicrob. Agents Chemother.* **46:**2365–2372.

Boston Interhospital Virus Study Group and the NIAID-Sponsored Cooperative Antiviral Clinical Study (1975). Failure of high dose 5-iodo-2'-deoxyuridine in the therapy of herpes simplex virus encephalitis . *N. Engl. J. Med.* **292:**599–603.

Bray, M., Driscoll, J., and Huggins, J. W. (2000). Treatment of lethal Ebola virus infection in mice with a single dose of an S-adenosyl-L-homocysteine hydrolase inhibitor. *Antiviral Res.* **45:**135–147.

Bray, M., Raymond, J. L., Geisbert, T., and Baker, R. O. (2002). 3-deazaneplanocin A induces massively increased interferon-alpha production in Ebola virus-infected mice. *Antiviral Res.* **55:**151–159.

Breeden, C. J., Hall, T. C., and Tyler, H. R. (1966). Herpes simplex encephalitis treated with systemic 5-iodo-2'-deoxyuridine. *Ann. Intern. Med.* **65:**1050–1056.

Broder, S., Yarchoan, R., Collins, J. M., Lane, H. C., Markham, P. D., Klecker, R. W., Redfield, R. R., Mitsuya, H., Hoth, D. F., and Gelmann, E. (1985). Effects of suramin on HTLV-III/LAV infection presenting as Kaposi's sarcoma or AIDS-related complex: Clinical pharmacology and suppression of virus replication *in vivo*. *Lancet* **ii:**627–630.

Brown, D. M., Todd, A. R., and Varadarajan, S. (1956). Nucleotides. Part XXXVII. The structure of uridylic acids a and b, and a synthesis of spongouridine (3-β-D-arabofuranosyluracil). *J. Chem. Soc.* 2388–2393.

Chou, S., and Marousek, G. I. (2008). Accelerated evolution of maribavir resistance in a cytomegalovirus exonuclease domain II mutant. *J. Virol.* **82:**246–253.

Chou, S., Wechel, L. C., and Marousek, G. I. (2007). Cytomegalovirus UL97 kinase mutations that confer maribavir resistance. *J. Infect. Dis.* **196:**91–94.

Cohen, S. S. (1960). Introduction to the biochemistry of D-arabinosyl nucleosides. *Prog. Nucleic Acid Res. Mol. Biol.* **5:**1–88.

Cooper, D. A., Steigbigel, R. T., Gatell, J. M., Rockstroh, J. K., Katlama, C., Yeni, P., Lazzarin, A., Clotet, B., Kumar, P. N., Eron, J. E., Schechter, M., Markowitz, M., *et al.* (2008). Subgroup and resistance analyses of raltegravir for resistant HIV-1 infection. *N. Engl. J. Med.* **359:**355–365.

Davies, W. L., Grunert, R. R., Haff, R. F., McGahen, J. W., Neumayer, E. M., Paulshock, M., Watts, J. C., Wood, T. R., Hermann, E. C., and Hoffmann, C. E. (1964). Antiviral activity of 1-adamantanamine (amantadine). *Science* **144:**862–863.

De Clercq, E. (1979). Suramin: A potent inhibitor of the reverse transcriptase of RNA tumor viruses. *Cancer Lett.* **8:**9–22.

De Clercq, E. (1980). Antiviral and antitumor activities of 5-substituted 2'-deoxyuridines. *Methods Find. Exp. Clin. Pharmacol.* **2:**253–267.

De Clercq, E. (1985). Antiviral and antimetabolic activities of neplanocins. *Antimicrob. Agents Chemother.* **28:**84–89.

De Clercq, E. (1987). S-adenosylhomocysteine hydrolase inhibitors as broad-spectrum antiviral agents. *Biochem. Pharmacol.* **36:**2567–2575.

De Clercq, E. (1997). In search of a selective antiviral chemotherapy. *Clin. Microbiol. Rev.* **10:**674–693.

De Clercq, E. (2001). Vaccinia virus inhibitors as a paradigm for the chemotherapy of poxvirus infections. *Clin. Microbiol. Rev.* **14:**382–397.

De Clercq, E. (2002). Cidofovir in the treatment of poxvirus infections. *Antiviral Res.* **55:**1–13.

De Clercq, E. (2003). New inhibitors of human cytomegalovirus (HCMV) on the horizon. *J. Antimicrob. Chemother.* **51:**1079–1083.

De Clercq, E. (2004a). Antiviral drugs in current clinical use. *J. Clin. Virol.* **30:**115–133.

De Clercq, E. (2004b). Antivirals and antiviral strategies. *Nat. Rev. Microbiol.* **2:**704–720.

De Clercq, E. (2004c). Non-nucleoside reverse transcriptase inhibitors (NNRTIs): Past, present and future. *Chem. Biodivers.* **1:**44–64.

De Clercq, E. (2006a). Antiviral agents active against influenza A viruses. *Nat. Rev. Drug Discov.* **5:**1015–1025.

De Clercq, E. (2006b). From adefovir to Atripla™ via tenofovir, Viread™ and Truvada™. *Future Virol.* **1:**709–715.

De Clercq, E. (2007). Acyclic nucleoside phosphonates: Past, present and future. Bridging chemistry to HIV, HBV, HCV, HPV, adeno-, herpes-, and poxvirus infections: The phosphonate bridge. *Biochem. Pharmacol.* **73:**911–922.

De Clercq, E. (2008a). Emerging antiviral drugs. *Expert Opin. Emerg. Drugs* **13:**393–416.

De Clercq, E. (2008b). The discovery of antiviral agents: Ten different compounds, ten different stories. *Med. Res. Rev.* **28:**929–953.

De Clercq, E. (2008c). Antivirals: Current state of the art. *Future Virology* **3:**393–405.

De Clercq, E. (2009a). Antiviral drug discovery: Ten more compounds, and ten more stories (Part B). *Med. Res. Rev.* **29:**571–610.

De Clercq, E. (2009b). Anti-HIV drugs: 25 compounds approved within 25years after the discovery of HIV. *Int. J. Antimicrob. Agents* **33**:307–320.

De Clercq, E., and Field, H. J. (2006). Antiviral prodrugs—the development of successful prodrug strategies for antiviral chemotherapy. *Br. J. Pharmacol.* **147**:1–11.

De Clercq, E., and Field, H. J. (2008a). Antiviral Chemistry & Chemotherapy's current antiviral agents FactFile (2nd edition): RNA viruses. *Antiviral Chem. Chemother.* **19**:63–74.

De Clercq, E., and Field, H. J. (2008b). Antiviral Chemistry & Chemotherapy's current antiviral agents FactFile (2nd edition): Retroviruses and hepadnaviruses. *Antiviral Chem. Chemother.* **19**:75–105.

De Clercq, E., and Holý, A. (2005). Acyclic nucleoside phosphonates: A key class of antiviral drugs. *Nat. Rev. Drug Discov.* **4**:928–940.

De Clercq, E., and Montgomery, J. A. (1983). Broad-spectrum antiviral activity of the carbocyclic analog of 3-deazaadenosine. *Antiviral Res.* **3**:17–24.

De Clercq, E., Descamps, J., De Somer, P., and Holý, A. (1978). (S)-9-(2, 3-Dihydroxypropyl)adenine: An aliphatic nucleoside analog with broad-spectrum antiviral activity. *Science* **200**:563–565.

De Clercq, E., Descamps, J., De Somer, P., Barr, P. J., Jones, A. S., and Walker, R. T. (1979). (E)-5-(2-Bromovinyl)-2'-deoxyuridine: A potent and selective anti-herpes agent. *Proc. Natl. Acad. Sci. USA* **76**:2947–2951.

De Clercq, E., Degreef, H., Wildiers, J., de Jonge, G., Drochmans, A., Descamps, J., and De Somer, P. (1980). Oral (E)-5-(2-bromovinyl)-2'-deoxyuridine in severe herpes zoster. *Br. Med. J.* **281**:1178.

De Clercq, E., Holý, A., Rosenberg, I., Sakuma, T., Balzarini, J., and Maudgal, P. C. (1986). A novel selective broad-spectrum anti-DNA virus agent. *Nature* **323**:464–467.

De Clercq, E., Sakuma, T., Baba, M., Pauwels, R., Balzarini, J., Rosenberg, I., and Holý, A. (1987). Antiviral activity of phosphonylmethoxyalkyl derivatives of purine and pyrimidines. *Antiviral Res.* **8**:261–272.

De Clercq, E., Cools, M., Balzarini, J., Marquez, V. E., Borcherding, D. R., Borchardt, R. T., Drach, J. C., Kitaoka, S., and Konno, T. (1989). Broad-spectrum antiviral activities of neplanocin A, 3-deazaneplanocin A, and their 5'-nor derivatives. *Antimicrob. Agents Chemother.* **33**:1291–1297.

De Clercq, E., Andrei, G., Snoeck, R., De Bolle, L., Naesens, L., Degrève, B., Balzarini, J., Zhang, Y., Schols, D., Leyssen, P., Ying, C., and Neyts, J. (2001). Acyclic/carbocyclic guanosine analogues as anti-herpesvirus agents. *Nucleosides Nucleotides Nucleic Acids* **20**:271–285.

Deng, L., Dai, P., Ciro, A., Smee, D. F., Djaballah, H., and Shuman, S. (2007). Identification of novel antipoxviral agents: Mitoxantrone inhibits vaccinia virus replication by blocking virion assembly. *J. Virol.* **81**:13392–13402.

De Palma, A. M., Vliegen, I., De Clercq, E., and Neyts, J. (2008). Selective inhibitors of picornavirus replication. *Med. Res. Rev.* **28**:823–884.

Domagk, G., Behnisch, R., Mietzch, F., and Schmidt, H. (1946). Über eine neue, gegen Tuberkelbazillen *in vitro* wirksame Verbindungsklasse. *Naturwissenschaften* **10**:315.

Dorsey, B. D., and Vacca, J. P. (2001). Discovery and early development of indinavir. *In* "Protease Inhibitors in AIDS Therapy" (R. C. Ogden and C. W. Flexner, eds.), pp. 65–83. Informa Healthcare.

Duncan, I. B., and Redshaw, S. (2001). Discovery and early development of saquinavir. *In* "Protease Inhibitors in AIDS Therapy" (R. C. Ogden and C. W. Flexner, eds.), pp. 27–47. Informa Healthcare.

Eaton, M. D., Cheever, F. S., and Levenson, C. G. (1951). Further observations of the effect of acridines on the growth of viruses. *J. Immunol.* **66**:463–476.

Eaton, M. D., Perry, M. E., Levenson, C. G., and Gocke, I. M. (1952). Studies on the mode of action of aromatic diamidines of influenza and mumps virus in tissue culture. *J. Immunol.* **68**:321–334.

Eggers, H. J. (1976). Successful treatment of enterovirus-infected mice by 2-(alpha-hydroxybenzyl)-benzimidazole and guanidine. *J. Exp. Med.* **143**:1367–1381.

Eggers, H. J., and Tamm, I. (1961). Spectrum and characteristics of the virus inhibitory action of 2-(α-hydroxybenzyl)-benzimidazole. *J. Exp. Med.* **113**:657–682.

Eggers, H. J., and Tamm, I. (1962). On the mechanism of selective inhibition of enterovirus multiplication by 2-(α-hydroxybenzyl)-benzimidazole. *Virology* **18**:426–438.

Eggers, H. J., and Tamm, I. (1963). Inhibition of enterovirus ribonucleic acid synthesis by 2-(α-hydroxybenzyl)-benzimidazole. *Nature* **197**:1327–1328.

Elion, G. B., Furman, P. A., Fyfe, J. A., de Miranda, P., Beauchamp, L., and Schaeffer, H. J. (1977). Selectivity of action of an antiherpetic agent, 9-(2-hydroxyethoxymethyl)guanine. *Proc. Natl. Acad. Sci. USA* **74**:5716–5720.

Erickson, J. W. (2001). HIV-1 protease as a target for AIDS therapy. *In* "Protease Inhibitors in AIDS Therapy" (R. C. Ogden and C. W. Flexner, eds.), pp. 1–25. Informa Healthcare.

Field, H. J., and De Clercq, E. (2004). Antiviral drugs—a short history of their discovery and development. *Microbiol. Today* **31**:58–61.

Field, H. J., and De Clercq, E. (2008). Antiviral Chemistry & Chemotherapy's current antiviral agents FactFile (2nd edition): DNA viruses. *Antiviral Chem. Chemother.* **19**:51–62.

Fleming, A. (1929). On the antibacterial action of cultures of a penicillum, with special reference to their use in the isolation of *B. influenzae*. *Br. J. Exp. Pathol.* **10**:226–236.

Fox, J. J., Yung, N., and Bendich, A. (1957). Pyrimidine nucleosides. II. The synthesis of 1-β-D-arabinofuranosylthymine ("spongothymidine") *J. Am. Chem. Soc.* **79**:2775–2778.

Gallant, J. E., DeJesus, E., Arribas, J. R., Pozniak, A. L., Gazzard, B., Campo, R. E., Lu, B., McColl, D., Chuck, S., Enejosa, J., Toole, J. J., and Cheng, A. K. (2006). Tenofovir DF, emtricitabine, and efavirenz vs zidovudine, lamivudine, and efavirenz for HIV. *N. Engl. J. Med.* **354**:251–260.

Gauri, K. K. (1968). Subconjunctival application of 5-ethyl-2'-desoxyuridine (EDU) for the chemotherapy of experimental herpetic keratitis in the rabbit. *Klin. Monatsbl. Augenheilkd.* **153**:837–840.

Gauri, K. K., and Malorny, G. (1967). Chemotherapie der Herpes-Infektion mit neuen 5-Alkyluracildesoxyribosiden. *Naunyn Schmiedebergs Arch. Pharmakol. Exp. Pathol.* **257**:21–22.

Graves, M. C., Lim, J. J., Heimer, E. P., and Kramer, R. A. (1988). An 11-kDa form of human immunodeficiency virus protease expressed in *Escherichia coli* is sufficient for enzymatic activity. *Proc. Natl. Acad. Sci. USA* **85**:2449–3453.

Greenlee, W. J. (1990). Renin inhibitors. *Med. Res. Rev.* **10**:173–236.

Grinsztejn, B., Nguyen, B. Y., Katlama, C., Gatell, J. M., Lazzarin, A., Vittecoq, D., Gonzalez, C. J., Chen, J., Harvey, C. M., and Isaacs, R. D. (2007). Safety and efficacy of the HIV-1 integrase inhibitor raltegravir (MK-0518) in treatment-experienced patients with multidrug-resistant virus: A phase II randomised controlled trial. *Lancet* **369**:1261–1269.

Hadziyannis, S. J., Tassopoulos, N. C., Heathcote, E. J., Chang, T. T., Kitis, G., Rizzetto, M., Marcellin, P., Lim, S. G., Goodman, Z., Wulfsohn, M. S., Xiong, S., Fry, J., *et al.* (2003). Adefovir dipivoxil for the treatment of hepatitis B e antigen-negative chronic hepatitis B. *N. Engl. J. Med.* **348**:800–807.

Hadziyannis, S. J., Tassopoulos, N. C., Heahtcote, E. J., Chang, T. T., Kitis, G., Rizzetto, M., Marcellin, P., Lim, S. G., Goodman, Z., Ma, J., Arterburn, S., Xiong, S., et al. (2005). Long-term therapy with adefovir dipivoxil for HBeAg-negative chronic hepatitis B. N. Engl. J. Med. **352:**2673–2681.

Hamre, D., Bernstein, J., and Donovick, R. (1950). Activity of p-aminobenzalde-hyde, 3-thiosemicarbazone on vaccinia virus in the chick embryo and in the mouse. Proc. Soc. Exp. Biol. Med. **73:**275–278.

Hamre, D., Brownlee, K. A., and Donovick, R. (1951). Studies on the chemotherapy of vaccinia virus. II. The activity of some thiosemicarbazones. J. Immunol. **67:**305–312.

Hazuda, D. J., Felock, P., Witmer, M., Wolfe, A., Stillmock, K., Grobler, J. A., Espeseth, A., Gabryelski, L., Schleif, W., Blau, C., and Miller, M. D. (2000). Inhibitors of strand transfer that prevent integration and inhibit HIV-1 replica-tion in cells. Science **287:**646–650.

Hazuda, D. J., Young, S. D., Guare, J. P., Anthony, N. J., Gomez, R. P., Wai, J. S., Vacca, J. P., Handt, L., Motzel, S. L., Klein, H. J., Dornadula, G., Danovich, R. M., et al. (2004). Integrase inhibitors and cellular immunity sup-press retroviral replication in rhesus macaques. Science **305:**528–532.

He, G., Qiao, J., Dong, C., He, C., Zhao, L., and Tian, Y. (2008). Amantadine-resistance among H5N1 avian influenza viruses isolated in Northern China. Antiviral Res. **77:**72–76.

Herrmann, E. C., Jr. (1961). Plaque inhibition test for detection of specific inhibi-tors of DNA containing viruses. Proc. Soc. Exp. Biol. Med. **107:**142–145.

Hollingshead, A. C., and Smith, P. K. (1958). Effects of certain purines and related compounds on virus propagation. J. Pharmacol. Exp. Ther. **123:**54–62.

Horsfall, F. L., Jr. (1955). Approaches to the chemotherapy of viral diseases. Bull. N.Y. Acad. Med. **31:**783–793.

Horsfall, F. L., Jr., and McCarty, M. (1947). The modifying effects of certain substances of bacterial origin on the course of infection with pneumonia virus of mice (PVM). J. Exp. Med. **85:**623–645.

Horsfall, F. L., Jr., and Tamm, I. (1957). Chemotherapy of viral and rickettsial diseases. Annu. Rev. Microbiol. **11:**339–370.

Huggins, J., Zhang, Z. X., and Bray, M. (1999). Antiviral drug therapy of filovirus infections: S-adenosylhomocysteine hydrolase inhibitors inhibit Ebola virus in vitro and in a lethal mouse model. J. Infect. Dis. **179**(Suppl. 1):S240–S247.

Hurst, E. W., Melvin, P., and Peters, J. M. (1952). The prevention of encephalitis due to the viruses of Eastern equine encephalomyelitis and louping-ill: Experi-ments with trypan red, mepacrine, and many other substances. Br. J. Pharmacol. **7:**455–472.

Isaacs, A., and Lindenmann, J. (1957). Virus interference. I. The interferon. Proc. R. Soc. Lond. B. Biol. Sci. **147:**258–267.

Janssen, P. A., Lewi, P. J., Arnold, E., Daeyaert, F., de Jonge, M., Heeres, J., Koymans, L., Vinkers, M., Guillemont, J., Pasquier, E., Kukla, M., Ludovici, D., et al. (2005). In search of a novel anti-HIV drug: Multidisciplinary coordination in the discovery of 4-[[4-[[4-[(1E)-2-cyanoethenyl]-2, 6-dimethyl-phenyl]amino]-2-pyrimidinyl]amino]-benzonitrile (R278474, rilpivirine). J. Med. Chem. **48:**1901–1909.

Jungeblut, C. W. (1951). Chemotherapeutic effects of a naphthoquinonimine on infection of mice with Columbia-SK group of viruses. Proc. Soc. Exp. Biol. Med. **77:**176–182.

Kaufman, H. E. (1962). Clinical cure of herpes simplex keratitis by 5-iodo-2′-deoxyuridine. Proc. Soc. Exp. Biol. Med. **109:**251–252.

Kaufman, H. E., and Heidelberger, C. (1964). Therapeutic antiviral action of 5-trifluoromethyl-2'-deoxyuridine in herpes simplex keratitis. *Science* **145:**585–586.

Kaufman, H. E., and Maloney, E. D. (1962a). IDU and hydrocortisone in experimental herpes simplex keratitis. *Arch. Ophthalmol.* **68:**396–398.

Kaufman, H. E., Nesburn, A. B., and Maloney, E. D. (1962a). IDU therapy of herpes simplex. *Arch. Ophthalmol.* **67:**583–591.

Kaufman, H. E., Martola, E.-L., and Dohlman, C. (1962b). Use of 5-iodo-2'-deoxyuridine (IDU) in treatment of herpes simplex keratitis. *Arch. Ophthalmol.* **68:**235–239.

Kaufman, H. E., Martola, E. L., and Dohlman, C. H. (1963). Herpes simplex treatment with IDU and corticosteroids. *Arch. Ophthalmol.* **69:**468–472.

Kempf, D. J. (2001). Discovery and early development of ritonavir. *In* "Protease Inhibitors in AIDS Therapy" (R. C. Ogden and C. W. Flexner, eds.), pp. 49–64. Informa Healthcare.

Kim, C. U., Lew, W., Williams, M. A., Liu, H., Zhang, L., Swaminathan, S., Bischofberger, N., Chen, M. S., Mendel, D. B., Tai, C. Y., Laver, G., and Stevens, R. C. (1997). Influenza neuraminidase inhibitors possessing a novel hydrophobic interaction in the enzyme active site: Design, synthesis, and structural analysis of carbocyclic sialic acid analogues with potent anti-influenza activity. *J. Am. Chem. Soc.* **119:**681–690.

Kissman, H. M., Child, R. G., and Weiss, M. J. (1957). Synthesis and biological properties of certain 5, 6-dichlorobenzimidazole ribosides. *J. Am. Chem. Soc.* **79:**1185–1188.

Lagoja, I. M., and De Clercq, E. (2008). Anti-influenza virus agents: Synthesis and mode of action. *Med. Res. Rev.* **28:**1–38.

Lee, W. W., Benitez, A., Goodman, L., and Baker, B. R. (1960). Potential anticancer agents. XL. Synthesis of the β-anomer of 9-(β-D-arabinofuranosyl)-adenine. *J. Am. Chem. Soc.* **82:**2648–2649.

Leyssen, P., De Clercq, E., and Neyts, J. (2008). Molecular strategies to inhibit the replication of RNA viruses. *Antiviral Res.* **78:**9–25.

Manns, M. P., Foster, G. R., Rockstroh, J. K., Zeuzem, S., Zoulim, F., and Houghton, M. (2007). The way forward in HCV treatment-finding the right path. *Nat. Rev. Drug Discov.* **6:**991–1000.

Marcellin, P., Chang, T. T., Lim, S. G., Tong, M. J., Sievert, W., Shiffman, M. L., Jeffers, L., Goodman, Z., Wulfsohn, M. S., Xiong, S., Fry, J., and Brosgart, C. L. (2003). Adefovir dipivoxil for the treatment of hepatitis B e antigen-positive chronic hepatitis B. *N. Engl. J. Med.* **348:**808–816.

Marcellin, P., Heathcote, E. J., Buti, M., Gane, E., de Man, R. A., Krastev, Z., Germanidis, G., Lee, S. S., Flisiak, R., Kaita, K., Manns, M., Kotzev, I., *et al.* (2008). Tenofovir disoproxil fumarate versus adefovir dipivoxil for chronic hepatitis B. *N. Engl. J. Med.* **359:**2442–2455.

Matthews, T., Salgo, M., Greenberg, M., Chung, J., DeMasi, R., and Bolognesi, D. (2004). Enfuvirtide: The first therapy to inhibit the entry of HIV-1 into host CD4 lymphocytes. *Nat. Rev. Drug Discov.* **3:**215–225.

Mitsuya, H., and Broder, S. (1986). Inhibition of the *in vitro* infectivity and cytopathic effect of human T-lymphotrophic virus type III/lymphadenopathy-associated virus (HTLV-III/LAV) by 2',3'-dideoxynucleosides. *Proc. Natl. Acad. Sci. USA* **83:**1911–1915.

Mitsuya, H., Popovic, M., Yarchoan, R., Matsushita, S., Gallo, R. C., and Broder, S. (1984). Suramin protection of T cells *in vitro* against infectivity and cytopathic effect of HTLV-III. *Science* **226:**172–174.

Mitsuya, H., Weinhold, K. J., Furman, P. A., St. Clair, M. H., Lehrman, S. N., Gallo, R. C., Bolognesi, D., Barry, D. W., and Broder, S. (1985). 3'-Azido-3'-deoxythymidine (BW A509U): An antiviral agent that inhibits the infectivity and cytopathic effect of human T-lymphotropic virus type III/lymphadenopathy-associated virus *in vitro*. *Proc. Natl. Acad. Sci. USA* **82:**7096–7100.

Miyasaka, T., Tanaka, H., Baba, M., Hayakawa, H., Walker, R. T., Balzarini, J., and De Clercq, E. (1989). A novel lead for specific anti-HIV-1 agents: 1-[(2-hydroxyethoxy)methyl]-6-(phenylthio)thymine. *J. Med. Chem.* **32:**2507–2509.

Moore, A. E., and Friend, C. (1951). Effect of 2, 6-diaminopurine on the course of Russian spring-summer encephalitis infection in the mouse. *Proc. Soc. Exp. Biol. Med.* **78:**153–157.

Mous, J., Heimer, E. P., and Le Grice, S. F. (1988). Processing protease and reverse transcriptase from human immunodeficiency virus type I polyprotein in *Escherichia coli. J. Virol.* **62:**1433–1436.

Naesens, L., Bischofberger, N., Augustijns, P., Annaert, P., Van den Mooter, G., Arimilli, M. N., Kim, C. U., and De Clercq, E. (1998). Antiretroviral efficacy and pharmacokinetics of oral bis(isopropyloxycarbonyloxymethyl)-9-(2-phosphonylmethoxypropyl)adenine in mice. *Antimicrob. Agents Chemother.* **42:**1568–1573.

Nemes, M. M., and Hilleman, M. R. (1965). Effective treatment of experimental herpes simplex keratitis with new derivtive, 5-methylamino-2'-deoxyuridine (MADU). *Proc. Soc. Exp. Biol. Med.* **119:**515–520.

Neyts, J., Leyssen, P., Verbeken, E., and De Clercq, E. (2004). Efficacy of cidofovir in a murine model of disseminated progressive vaccinia. *Antimicrob. Agents Chemother.* **48:**2267–2273.

Nolan, D. C., Carruthers, M. M., and Lerner, A. M. (1970). Herpesvirus hominis encephalitis in Michigan. Report of thirteen cases, including six treated with idoxuridine. *N. Engl. J. Med.* **282:**10–13.

Nolan, D. C., Lauter, C. B., and Lerner, M. (1973). Idoxuiridne in herpes simplex virus (type 1) encephalitis. *Ann. Intern. Med.* **78:**243–246.

Otten, R. A., Smith, D. K., Adams, D. R., Pullium, J. K., Jackson, E., Kim, C. N., Jaffe, H., Janssen, R., Butera, S., and Folks, T. M. (2000). Efficacy of postexposure prophylaxis after intravaginal exposure of pig-tailed macaques to a human-derived retrovirus (human immunodeficiency virus type 2). *J. Virol.* **74:**9771–9775.

Pauwels, R. (2006). Aspects of successful drug discovery and development. *Antiviral Res.* **71:**77–89.

Pauwels, R., Andries, K., Desmyter, J., Schols, D., Kukla, M. J., Breslin, H. J., Raeymaeckers, A., Van Gelder, J., Woestenborghs, R., Heykants, J., Schellekens, K., Janssen, M. A. C., *et al.* (1990). Potent and selective inhibition of HIV-1 replication *in vitro* by a novel series of TIBO derivatives. *Nature* **343:**470–474.

Perros, M. (2007). CCR5 antagonists for the treatment of HIV infection and AIDS. *Adv. Antiviral Drug Des.* **5:**185–212.

Pommier, Y., Johnson, A. A., and Marchand, C. (2005). Integrase inhibitors to treat HIV/AIDS. *Nat. Rev. Drug Discov.* **4:**2236–2248.

Potter, C. W., Phair, J. P., Vodinelich, L., Fenton, R., and Jennings, R. (1976). Antiviral immunosuppressive and antitumor effects of ribavirin. *Nature* **259:**496–497.

Pozniak, A. L., Gallant, J. E., DeJesus, E., Arribas, J. R., Gazzard, B., Campo, R. E., Chen, S. S., McColl, D., Enejosa, J., Toole, J. J., and Cheng, A. K. (2006). Tenofovir disoproxil fumarate, emtricitabine, and efavirenz versus fixed-dose zidovudine/lamivudine and efavirenz in antiretroviral-naive patients: Virologic, immunologic, and morphologic changes—a 96-week analysis. *J. Acquir. Immune Defic. Syndr.* **43:**535–540.

Privat de Garilhe, M., and de Rudder, J. (1964). Effet de deux nucléosides de l'arabinose sur la multiplication des virus de l'herpès et de la vaccine en culture cellulaire. *C. R. Acad. Sci. Paris* **259:**2725–2728.

Prusoff, W. H. (1959). Synthesis and biological activities of iododeoxyuridine, an analog of thymidine. *Biochim. Biophys. Acta* **32:**295–296.

Prusoff, W. H. (1960). Studies on the mechanism of action of 5-iododeoxyuridine, an analog of thymidine. *Cancer Res.* **20:**92–95.

Prusoff, W. H., Holmes, W. L., and Welch, A. D. (1953). Non-utilization of radioactive iodinated uracil, uridine, and orotic acid by animal tissues *in vivo. Cancer Res.* **13:**221–225.

Quenelle, D. C., Buller, R. M., Parker, S., Keith, K. A., Hruby, D. E., Jordan, R., and Kern, E. R. (2007a). Efficacy of delayed treatment with ST-246 given orally against systemic orthopoxvirus infections in mice. *Antimicrob. Agents Chemother.* **51:**689–695.

Quenelle, D. C., Prichard, M. N., Keith, K. A., Hruby, D. E., Jordan, R., Painter, G. R., Robertson, A., and Kern, E. R. (2007b). Synergistic efficacy of the combination of ST-246 with CMX001 against orthopoxviruses. *Antimicrob. Agents Chemother.* **51:**4118–4124.

Reeves, P. M., Bommarious, B., Lebeis, S., McNulty, S., Christensen, J., Swimm, A., Chahroudi, A., Chavan, R., Feinberg, M. B., Veach, D., Bornmann, W., Sherman, M., *et al.* (2005). Disabling poxvirus pathogenesis by inhibition of Abl-family tyrosine kinases. *Nat. Med.* **11:**731–739.

Robbins, B. L., Srinivas, R. V., Kim, C., Bischofberger, N., and Fridland, A. (1998). Anti-human immunodeficiency virus activity and cellular metabolism of a potential prodrug of the acyclic nucleoside phosphonate 9-*R*-(2-phosphonomethoxypropyl)adenine (PMPA), bis(isopropyloxymethylcarbonyl)-PMPA. *Antimicrob. Agents Chemother.* **42:**612–617.

Roberts, N. A., Martin, J. A., Kinchington, D., Broadhurst, A. V., Craig, J. C., Duncan, I. B., Galpin, S. A., Handa, B. K., Kay, J., and Kröhn, A. (1990). Rational design of peptide-based HIV proteinase inhibitors. *Science* **248:**358–361.

Sbrana, E., Jordan, R., Hruby, D. E., Mateo, R. I., Xiao, S. Y., Siirin, M., Newman, P. C., Da Rosa, A. P., and Tesh, R. B. (2007). Efficacy of the antipoxvirus compound ST-246 for treatment of severe orthopoxvirus infection. *Am. J. Trop. Med. Hyg.* **76:**768–773.

Schabel, F. M., Jr. (1968). The antiviral activity of 9-β-D-arabinofuranosyladenine (Ara-A). *Chemotherapy* **13:**321–338.

Schabel, F. M., Jr., and Montgomery, J. A. (1972). Purines and pyrimidines. *In* "Chemotherapy of Virus Diseases" (D. J. Bauer, ed.), Vol. 1, pp. 231–363. Pergamon Press, Oxford.

Schaeffer, H. J., Beauchamp, L., de Miranda, P., Elion, G. B., Bauer, D. J., and Collins, P. (1978). 9-(2-hydroxyethoxymethyl) guanine activity against viruses of the herpes group. *Nature* **272:**583–585.

Sidwell, R. W., Huffman, J. H., Khare, G. P., Allen, L. B., Witkowski, J. T., and Robins, R. K. (1972). Broad-spectrum antiviral activity of virazole: 1-β-D-ribofuranosyl-1, 2, 4-triazole-3-carboxamide. *Science* **177:**705–706.

Smee, D. F., and Sidwell, R. W. (2003). A review of compounds exhibiting antiorthopoxvirus activity in animal models. *Antiviral Res.* **57:**41–52.

Smee, D. F., Hurst, B. L., Wong, M.-H., Glazer, R. I., Rahman, A., and Sidwell, R. W. (2007). Efficacy of *N*-methanocarbathymidine in treating mice infected intranasally with the IHD and WR strains of vaccinia virus. *Antiviral Res.* **76:**124–129.

Steigbigel, R. T., Cooper, D. A., Kumar, P. N., Eron, J. E., Schechter, M., Markowitz, M., Loutfy, M. R., Lennox, J. L., Gatell, J. M., Rockstroh, J. K., Katlama, C., Yeni, P., *et al.* (2008). Raltegravir with optimized background therapy for resistant HIV-1 infection. *N. Engl. J. Med.* **359:**339–354.

Streeter, D. G., Witkowski, J. T., Khare, G. P., Sidwell, R. W., Bauer, R. J., Robins, R. K., and Simon, L. N. (1973). Mechanism of action of 1-β-D-ribofur-anosyl-1, 2, 4-triazole-3-carboxamide (Virazole), a new broad-spectrum antiviral agent. *Proc. Natl. Acad. Sci. USA* **70:**1174–1178.

Swierkowski, M., and Shugar, D. (1969). A nonmutagenic thymidine analog with antiviral activity. 5-Ethyldeoxyuridine. *J. Med. Chem.* **12:**533–534.

Tamm, I. (1956a). Antiviral chemotherapy. *Yale J. Biol. Med.* **29:**33–49.

Tamm, I. (1956b). Selective chemical inhibition of multiplication. *J. Bacteriol.* **72:**42–53.

Tamm, I. (1960). Symposium on the experimental pharmacology and clinical use of antimetabolites. Part III. Metabolic antagonists and selective virus inhibition. *Clin. Pharmacol. Ther.* **1:**777–796.

Tamm, I., and Nemes, M. M. (1957). Glycosides of chlorobenzimidazoles as inhibitors of poliovirus multiplication. *Virology* **4:**483–498.

Tamm, I., and Nemes, M. M. (1959). Selective inhibition of poliovirus multiplication. *J. Clin. Invest.* **38:**1047–1048.

Tamm, I., and Overman, J. R. (1957). Relationship between structure of benzimidazole derivatives and inhibitory activity on vaccinia virus multiplication. *Virology* **3:**185–196.

Tamm, I., and Tyrrell, D. A. J. (1954). Influenza virus multiplication in the chorioallantoic membrane *in vitro*: Kinetic aspects of inhibition by 5, 6-dichloro-1-β-D-ribofuranosyl-benzimidazole. *J. Exp. Med.* **100:**541–562.

Tamm, I., Folkers, K., and Horsfall, F. L., Jr. (1952). Inhibition of influenza virus multiplication by 2, 5-dimethylbenzimidazole. *Yale J. Biol. Med.* **24:**559–567.

Tamm, I., Folkers, K., Shunk, C. H., and Horsfall, F. L., Jr. (1954). Inhibition of influenza virus multiplication by N-glycosides of benzimidazoles. *J. Exp. Med.* **99:**227–250.

Tamm, I., Folkers, K., and Shunk, C. H. (1956). High inhibitory activity of certain halogenated ribofuranosylbenzimidazoles on influenza B virus multiplication. *J. Bacteriol.* **72:**54–58.

Tamm, I., Nemes, M. M., and Osterhout, S. (1960). On the role of ribonucleic acid in animal virus synthesis. I. Studies with 5, 6-dichloro-1-β-D-ribofuranosylbenzimidazole. *J. Exp. Med.* **111:**339–349.

Thompson, R. L. (1947). The effect of metabolites, metabolite antagonists and enzyme-inhibitors on the growth of the vaccinia virus in Maitland type of tissue cultures. *J. Immunol.* **55:**345–352.

Thompson, R. L., Wilkin, M. L., Hitchings, G. H., Elion, G. B., Falco, E. A., and Russell, P. B. (1949a). The effects of antagonists on the multiplication of vaccinia virus *in vitro*. *Science* **110:**454.

Thompson, R. L., Wilkin, M. L., Hitchings, G. H., and Russell, P. B. (1949b). The virostatic and virucidal action of α-haloacylamides on vaccinia virus *in vitro*. *Proc. Soc. Exp. Biol. Med.* **72:**169–171.

Thompson, R. L., Price, M., Minton, S. A., Jr., Falco, E. A., and Hitchings, G. H. (1951a). Protection of mice against the vaccinia virus by the administration of phenoxythiouracils. *J. Immunol.* **67:**483–491.

Thompson, R. L., Price, M. L., and Minton, S. A., Jr. (1951b). Protection of mice against vaccinia virus by administration of benzaldehyde thiosemicarbazone. *Proc. Soc. Exp. Biol. Med.* **78:**11–13.

Thompson, R. L., Minton, S. A., Jr., Officer, J. E., and Hitchings, G. H. (1953a). Effect of heterocyclic and other thiosemicarbazones on vaccinia infection in the mouse. *J. Immunol.* **70:**229–234.

Thompson, R. L., Davis, J., Russell, P. B., and Hitchings, G. H. (1953b). Effect of aliphatic oxime and isatin thiosemicarbazone on vaccinia infection in the mouse and in the rabbit. *Proc. Soc. Exp. Biol. Med.* **84:**496–499.

Tsai, C. C., Follis, K. E., Sabo, A., Beck, T. W., Grant, R. F., Bischofberger, N., Benveniste, R. E., and Black, R. (1995). Prevention of SIV infection in macaques by (R)-9-(2-phosphonylmethoxypropyl)adenine. *Science* **270:**1197–1199.

Van Rompay, K. K., McChesney, M. B., Aguirre, N. L., Schmidt, K. A., Bischofberger, N., and Marthas, M. L. (2001). Two low doses of tenofovir protect newborn macaques against oral simian immunodeficiency virus infection. *J. Infect. Dis.* **184:**429–438.

von Itzstein, M. (2007). The war against influenza: Discovery and development of sialidase inhibitors. *Nat. Rev. Drug Discov.* **6:**967–974.

von Itzstein, M., Wu, W. Y., Kok, G. B., Pegg, M. S., Dyason, J. C., Jin, B., Van Phan, T., Smythe, M. L., White, H. F., Oliver, S. W., Colman, P. M., Varghese, J. N., *et al.* (1993). Rational design of potent sialidase-based inhibitors of influenza virus replication. *Nature* **363:**418–423.

Vora, S., Damon, I., Fulginiti, V., Weber, S. G., Kahana, M., Stein, S. L., Gerber, S. I., Garcia-Houchins, S., Lederman, E., Hruby, D., Collins, L., Scott, D., *et al.* (2008). Severe eczema vaccinatum in a household contact of a smallpox vaccinee. *Clin. Infect. Dis.* **46:**1555–1561.

Votruba, I., Holý, A., and De Clercq, E. (1983). Metabolism of the broad-spectrum antiviral agent, 9-(S)-(2, 3-dihydroxypropyl)adenine, in different cell lines. *Acta Virol.* **27:**273–276.

Welch, A. D., and Prusoff, W. H. (1960). A synopsis of recent investigations of 5-iodo-2'-deoxyuridine. *Cancer Chemother. Rep.* **6:**29–36.

Whitley, R. J., Ch'ien, L. T., Dolin, R., Galasso, G. J., and Alford, C. A., Jr. (1976). Adenine arabinoside therapy of herpes zoster in the immunosuppressed. NIAID collaborative antiviral study. *N. Engl. J. Med.* **294:**1193–1199.

Whitley, R. J., Soong, S. J., Dolin, R., Galasso, G. J., Ch'ien, L. T., and Alford, C. A. (1977). Adenine arabinoside therapy of biopsy-proved herpes simplex encephalitiis. National Institute of Allergy and Infectious Diseases collaborative antiviral study. *N. Engl. J. Med.* **297:**289–294.

Whitley, R. J., Arvin, A., Prober, C., Burchett, S., Corey, L., Powell, D., Plotkin, S., Starr, S., Alford, C., and Connor, J. (1991). A controlled trial comparing vidarabine with acyclovir in neonatal herpes simplex virus infection. Infectious Diseases Collaborative Antiviral Study Group. *N. Engl. J. Med.* **324:**444–449.

Yang, H., Kim, S. K., Kim, M., Reche, P. A., Morehead, T. J., Damon, I. K., Welsh, R. M., and Reinherz, E. L. (2005a). Antiviral chemotherapy facilitates control of poxvirus infections through inhibition of cellular signal transduction. *J. Clin. Invest.* **115:**379–387.

Yang, G., Pevear, D. C., Davies, M. H., Collett, M. S., Bailey, T., Rippen, S., Barone, L., Burns, C., Rhodes, G., Tohan, S., Huggins, J. W., Baker, R. O., *et al.* (2005b). An orally bioavailable antipoxvirus compound (ST-246) inhibits extracellular virus formation and protects mice from lethal orthopoxvirus challenge. *J. Virol.* **79:**13139–13149.

CHAPTER 2

Use of Animal Models to Understand the Pandemic Potential of Highly Pathogenic Avian Influenza Viruses

Jessica A. Belser,*,† **Kristy J. Szretter,***,1
Jacqueline M. Katz,* and **Terrence M. Tumpey***

Contents

I. Introduction   56
II. Influenza A Virus Subtypes and Host Range   57
III. Avian Influenza A Virus in Humans   60
IV. Use of the Mouse Model to Study Influenza Virus
    Pathogenesis   63
    A. Mouse model for human influenza A virus
       pathogenesis   63
    B. Mouse model for H5N1 virus pathogenesis   65
    C. Mouse model for H7 virus pathogenesis   67
    D. Gene knockout mice in the study of avian
       influenza   68
    E. Tropism of avian influenza viruses   70
V. Use of the Ferret Model to Study Influenza
   Virus Pathogenesis   71
    A. Ferret model for human influenza A virus
       pathogenesis   71
    B. Ferret model for H5N1 virus pathogenesis   72
    C. Ferret model for H7 virus pathogenesis   73

* Influenza Division, National Center for Immunization and Respiratory Diseases, Centers for Disease
Control and Prevention, Atlanta, Georgia 30333, USA
† Department of Microbiology, Mount Sinai School of Medicine, New York, New York 10029, USA
1 Current address: School of Medicine, Washington University, Campus Box 8051, 660 S Euclid, St. Louis,
Missouri 63110, USA

Advances in Virus Research, Volume 73
ISSN 0065-3527, DOI: 10.1016/S0065-3527(09)73002-7

D. Transmissibility of avian influenza viruses 74
VI. Molecular Basis of Avian Influenza Pathogenesis 75
    A. Hemagglutinin cleavage site 76
    B. Surface glycoproteins (HA and NA) 77
    C. Polymerase complex 78
    D. PB2 protein 79
    E. PB1-F2 protein 81
    F. NS1 protein 81
VII. Conclusions 82
Acknowledgments 84
References 84

Abstract   It has been 40 years since the last influenza pandemic and it is generally considered that another could occur at any time. Recent introductions of influenza A viruses from avian sources into the human population have raised concerns that these viruses may be a source of a future pandemic strain. Therefore, there is a need to better understand the pathogenicity of avian influenza viruses for mammalian species so that we may be better able to predict the pandemic potential of such viruses and develop improved methods for their prevention and control. In this review, we describe the virulence of H5 and H7 avian influenza viruses in the mouse and ferret models. The use of these models is providing exciting new insights into the contribution of virus and host responses toward avian influenza viruses, virus tropism, and virus transmissibility. Identifying the role of individual viral gene products and mapping the molecular determinants that influence the severity of disease observed follow-ing avian influenza virus infection is dependent on the use of reliable animal models. As avian influenza viruses continue to cause human disease and death, animal pathogenesis studies identify avenues of investigation for novel preventative and therapeutic agents that could be effective in the event of a future pandemic.

I. INTRODUCTION

Influenza viruses continue to pose a major public health concern. Winter epidemics of influenza A viruses occur annually in temperate climates, and are likely made all the more prevalent due to the growing world population and increasingly rapid international transportation systems. Pandemic influenza has the capacity to cause severe disease and death on a global scale. Increasing numbers of human infection with highly patho-genic avian influenza (HPAI) viruses of multiple subtypes has identified a need to better understand the pandemic potential of this group of viruses.

Globally, widespread outbreaks of seasonal influenza are estimated to cause 250,000 to 500,000 deaths annually. Each year in the United States alone, on average, 5–20% of the population is infected with influenza, with approximately 36,000 deaths from complications of influenza virus infection; 90% of these deaths are in the elderly population aged >65 years (Thompson *et al.*, 2004). Influenza A virus infection begins in the nasal and tracheal airways, and can spread throughout the upper and lower respiratory tract. Clinical symptoms of an acute human influenza A virus infection can range from mild to severe and typically include fever, cough, headache, and malaise. In addition to seasonal epidemics of human influenza viruses, multiple subtypes of H5 and H7 avian influenza viruses circulating in domestic poultry have in total infected over 500 individuals in the last decade with an approximate 50% fatality rate (CDC, 2004b; Eurosurveillance Editorial Team, 2007; Fouchier *et al.*, 2004; Nguyen-Van-Tam *et al.*, 2006; Peiris *et al.*, 1999; Tweed *et al.*, 2004; WHO, 2009). Avian influenza viruses of high pathogenicity are usually associated with severe clinical illness in humans (Abdel-Ghafar *et al.*, 2008). The direct bird-to-human transmission of avian subtype viruses has raised concerns that these viruses may be a source of the next pandemic strain. Therefore, it is crucial to study the pathogenicity of these avian viruses in suitable animal models to better understand the genetic markers responsible for virulence and transmissibility. Such knowledge would enhance our ability to predict the pandemic potential of avian influenza virus strains and aid scientists to develop improved methods for their prevention and control. The topics included in this review highlight areas of active research into the understanding of the pathogenicity of avian influenza viruses in mammalian hosts and the molecular determinants that confer high virulence.

II. INFLUENZA A VIRUS SUBTYPES AND HOST RANGE

Influenza viruses are single-stranded, negative-sense, enveloped RNA viruses within the family Orthomyxoviridae. Multiple types of the virus exist; however, influenza type A viruses have a broad host range and thus differ from both type B and C viruses which are generally restricted to humans (Easterday, 1975). As such, influenza A viruses will be the focus of this review. The virus contains eight gene segments, coding for 11 known proteins (Chen *et al.*, 2001; Webster *et al.*, 1992). The major surface glycoproteins are the hemagglutinin (HA) and neuraminidase (NA), which form the basis of multiple serologically distinct influenza A virus subtypes. There are 16 HA and 9 NA subtypes of influenza A viruses known to circulate in wild aquatic birds, the natural reservoir of all influenza A viruses (Fouchier *et al.*, 2005; Olsen *et al.*, 2006; Rohm *et al.*, 1996;

Webster *et al.*, 1992). Most influenza viruses cause asymptomatic infection in aquatic birds, in which replication occurs primarily in the epithelium of the intestinal tract. Fecal samples from wild birds may contain high titers of virus, suggesting a mechanism for the spread of virus among avian species via fecal–oral transmission or transmission through fecal contamination of water (Webster *et al.*, 1978). Some viruses within the H5 and H7 subtypes have been associated with severe disease and mortality in avian populations, most often when introduced into domestic land-based birds (Capua and Alexander, 2004). Interspecies transmission of avian influenza viruses from wild bird reservoirs into domestic poultry and accidental transmission to mammals including transient emergence in whales, seals, cats, and other mammals has been documented (Vahlenkamp and Harder, 2006).

In humans, seasonal influenza viruses replicate primarily in the upper airway epithelium and are expelled in respiratory secretions when an individual coughs, sneezes, or speaks. Individuals become infected either through direct inhalation of large or small droplets containing viruses or by indirect contact with fomites on contaminated surfaces (Alford *et al.*, 1966; Bean *et al.*, 1982; Lidwell, 1974). Despite the variety of possible combinations, only three HA subtypes (H1, H2, H3) and two NA subtypes (N1, N2) have caused widespread, sustained disease in humans to date. This is most likely due, at least in part, to the ability of human influenza viruses (H1–H3 subtypes) to preferentially bind sialic acid in a defined type of linkage that is found at high amounts on the human respiratory tract epithelium (Gambaryan *et al.*, 1997; Yao *et al.*, 2008). Glycoconjugates containing terminal sialic acid serve as the cellular receptor for influenza viruses, which are found in one of two major linkage conformations, Neu5Acα(2,3)-Gal or Neu5Acα(2,6)-Gal (Skehel and Wiley, 2000). Receptor-binding preferences are generally species specific and this preference is associated with the cellular tropism of these viruses; human influenza subtypes preferentially infect nonciliated cells of the airway epithelium, which possess α(2–6)-linked sialic acids on their surface, while avian subtype viruses prefer to infect ciliated cells, which possess α(2–3)-linked sialic acids (Ito and Kawaoka, 2000; Matrosovich *et al.*, 2004; Rogers and Paulson, 1983; Skehel and Wiley, 2000). While α(2–6) linkages are generally more prevalent in the upper respiratory tract, studies have demonstrated the presence of α(2–3) sialic acid linkages in the lower respiratory tract of humans. The receptor distribution may help explain the viral attachment of the HPAI H5N1 viruses deep in the lung and severity of H5N1 viral pneumonia in humans (Nicholls *et al.*, 2007; Shinya *et al.*, 2006; Uiprasertkul *et al.*, 2005; van Riel *et al.*, 2006). It is generally believed that a switch in receptor-binding preference would be a necessary step for avian influenza viruses in the generation of a pandemic virus conferring efficient transmission among humans. However, the binding of influenza viruses to sialic acids and efficient transmission may not be only restricted by the particular α(2–3) or α(2–6) linkage but also by the complex glycan structural topology of the sialic acids (Chandrasekaran *et al.*, 2008; Srinivasan *et al.*, 2008).

Avian influenza viruses may acquire pandemic traits by one or more virologic mechanisms. Antigenic drift is the result of an accumulation of point mutations that yield amino acid substitutions in antigenic sites of viral HA or NA glycoproteins. This mechanism, used by seasonal influenza A viruses, allows the virus to escape neutralization from the host immune system, and can result in the emergence of viruses that cause annual epidemics of influenza (Scholtissek *et al.*, 1993; Webster *et al.*, 1992). Antigenic shift is the result of the emergence of a virus with a novel HA and/or NA within an immunologically naïve human population. This variation can arise from reassortment between avian and circulating human influenza viruses (Webster *et al.*, 1992). A novel avian–human reassortant influenza virus that has acquired human virus-like receptor-binding properties and causes disease in humans may cause a pandemic. Three such pandemics occurred in the twentieth century, in 1918 (H1N1), 1957 (H2N2), and 1968 (H3N2), with varying severity. The 1918 pandemic was by far the most devastating of the twentieth century pandemic strains, attributed with 20–50 million deaths worldwide. In comparison, the combined mortality following the 1957 and 1968 pandemics was reduced over 200-fold (Johnson and Mueller, 2002). While the genetic composition of the influenza A viruses responsible for the 1957 (H2N2) and 1968 (H3N2) human pandemics is largely known (Webster *et al.*, 1992), it remains uncertain whether the 1918 virus was the result of adaptation of an avian virus to humans or whether it exchanged genes through reassortment or another mechanism (Gibbs and Gibbs, 2006; Taubenberger *et al.*, 2005). Identification of viral sequence data of pre-1918 human influenza samples are needed to better understand the origin of the 1918 virus.

Such limited information on pandemic influenza viruses coupled with the viruses' ability to randomly mutate and recombine its RNA genome makes it impossible to predict which subtype will emerge as the next pandemic strain, when it will occur, or its severity. However, a current concern is that repeated transmission of avian H5 or H7 viruses from infected poultry to humans could increase the likelihood of emergence of an avian–human reassortant virus or an avian influenza virus that has acquired molecular changes needed for efficient and sustained virus transmission among humans. Therefore, additional studies are needed to fully understand the molecular correlates determining virulence and efficient transmission of H5 and H7 viruses in mammalian model systems. Moreover, mammalian pathogenesis data provide valuable information for the development of avian influenza vaccines in preclinical testing.

III. AVIAN INFLUENZA A VIRUS IN HUMANS

Volunteer studies found that humans supported only limited replication of low pathogenicity avian influenza (LPAI) viruses when inoculated at high doses, suggesting that avian influenza A viruses, in general, are not well-suited for efficient human infection (Beare and Webster, 1991). However, HPAI viruses, limited to the H5 and H7 subtypes, are responsible for the most severe avian influenza virus outbreaks in poultry and infections in humans (CDC, 2004a; Tran *et al.*, 2004). When HPAI viruses infect poultry, the virus is excreted in high titers from both the respiratory and the digestive tracts resulting in rapid spread through a population of susceptible hosts (Swayne and Suarez, 2000). HPAI is an extremely contagious multiorgan systemic disease of poultry and high titers of infectious virus can be detected in most visceral organs. Most cases of H5N1 human virus infection have occurred following handling of sick or dead poultry, or visiting live bird markets, roughly a week before the onset of illness (Abdel-Ghafar *et al.*, 2008; Mounts *et al.*, 1999). To date, avian influenza viruses of the H5, H7, and H9 subtypes have been associated with disease in humans, though not all of these infections resulted in severe human disease (Butt *et al.*, 2005; Fouchier *et al.*, 2004; Peiris *et al.*, 1999; WHO, 2009).

Prior to 1997, direct transmission of HPAI viruses to humans was not considered a major health risk. However, during that year, HPAI viruses of the H5N1 subtype caused outbreaks of disease in domestic poultry in Hong Kong that exhibited the ability to cause human respiratory infection and death, killing 6 of the 18 documented cases (Claas *et al.*, 1998; de Jong *et al.*, 1997; Subbarao *et al.*, 1998). This was the first documented influenza outbreak caused by a wholly avian virus directly transmitting to humans from infected poultry and causing death. Patients had primary viral pneumonia complicated by acute respiratory distress syndrome and multiple organ failure. Evidence for extrapulmonary replication of H5N1 virus in humans was inconclusive due to the paucity of postmortem tissues for study. Lymphopenia and cytokine dysregulation were observed and proposed to contribute to the increased severity of disease (To *et al.*, 2001; Yuen *et al.*, 1998). However, conclusions from serum cytokine data were also limited by the small number and poorly timed specimens available.

In early 2003, avian influenza H5N1 viruses again caused two documented human cases and one death in a single-family cluster (Peiris *et al.*, 2004). Of these two cases, one was more severe, presenting with fever, cough, bloody sputum, bone pain, lymphopenia, pulmonary dysfunction, and ultimately death (Peiris *et al.*, 2004). Since late 2003, the continued presence of H5N1 virus in Southeast Asia and later expansion to Europe and Africa has to date contributed to greater than 435 documented human cases and 260 deaths in 15 different countries (Abdel-Ghafar *et al.*, 2008;

WHO, 2009). Human cases from the 2004 outbreak in Vietnam and Thailand presented similarly to previous H5N1 human cases with other symptoms including bleeding nose, gums, gastrointestinal, and respiratory tracts (CDC, 2004a; Tran *et al.*, 2004). Atypical symptoms including gastrointestinal distress followed by systemic organ failure or acute encephalitis were observed with selected human H5N1 cases; both patients lacked respiratory involvement suggesting more extensive tissue tropism (Apisarnthanarak *et al.*, 2004; de Jong *et al.*, 2005). High viral load, viremia, and elevated cytokine and chemokine responses were additionally associated with fatal H5N1 human cases (de Jong *et al.*, 2006). The severe clinical symptoms observed during the most recent H5N1 outbreaks, greater rapid time-to-death, and case fatality rates of over 60% suggest that the recent H5N1 viruses are even more virulent than those infecting humans in 1997 (Abdel-Ghafar *et al.*, 2008). This enhancement of virulence was reflected in ferret pathogenesis studies where the 2004/2005 human H5N1 isolates were more virulent than the 1997 human H5N1 viruses (Maines *et al.*, 2005). The apparent increase in virulence, coupled with the high persistence of virus in the region, has led many public health agencies to consider this subtype the greatest contemporary pandemic threat.

Avian influenza viruses within the H7 subtype have also spread from infected poultry to infect humans. In fact, there has been an increase in the number of human cases of H7 exposure over the past decade and some viruses found in this subtype have demonstrated changes in receptor binding that potentially move them one step closer to a pandemic phenotype. Prior to 2003, H7 human infections were historically rare and largely due to laboratory or occupational exposure. Two noteworthy human cases of conjunctivitis occurred following exposure with H7 infected animals. In 1980, an H7N7 virus, A/Seal/Mass/1/80 was isolated from an individual following exposure to infected seals and, in 1996, a LPAI H7N7 virus, A/England/268/96, was isolated from an individual following exposure to ducks (Kurtz *et al.*, 1996; Webster *et al.*, 1981). In 2003, there was a widespread poultry outbreak of HPAI H7N7 virus in The Netherlands, which resulted in over 80 cases of human infection (Fouchier *et al.*, 2004; Koopmans *et al.*, 2004). The majority of human infections resulted in conjunctivitis, with a few individuals experiencing respiratory symptoms. A notable exception to this relatively mild illness was a single fatality due to acute respiratory distress syndrome (Fouchier *et al.*, 2004). Also in 2003, an individual in New York presented to hospital with a fever and cough; a LPAI H7N2 virus, A/NY/107/03 (NY/107), was subsequently isolated from a respiratory specimen (CDC, 2004b). The individual recovered from the respiratory illness, but the nature of his initial exposure to and infection with the avian virus remains unknown. A HPAI H7N3 outbreak in British Columbia, Canada, in 2004 resulted in two cases of human infection, both resulting in conjunctivitis (Hirst *et al.*, 2004; Tweed *et al.*, 2004). More recently, H7 viruses have resulted in cases of human infection in the United Kingdom. In 2006, LPAI H7N3 virus, first detected in a poultry flock in eastern England, was isolated from a poultry worker with conjunctivitis (Nguyen-Van-Tam *et al.*, 2006). In 2007, poultry infected with a LPAI H7N2 virus were sold from a small market in the United Kingdom and caused an outbreak that resulted in four confirmed cases of H7 human infection and 19 additional symptomatic but PCR negative individuals (Eurosurveillance Editorial Team, 2007). Additionally, one of the hospitalized cases reportedly involved a patient with neurological and gastrointestinal presentations, not respiratory disease. Individuals exposed to the virus reported both conjunctivitis and influenza-like illness; three of the individuals with confirmed H7 infection required hospitalization for 3–7 days (Dudley, 2008). Serologic evidence of additional H7 human infections has been reported from outbreaks in Virginia in 2002 and Italy in 2003 (CDC, 2004c; Puzelli *et al.*, 2005). The increased frequency of human infection with H7 viruses since 2003, in both Europe and North America, suggests that more research is needed to assess the pandemic potential of viruses found in this subtype.

The aforementioned clinical and epidemiological data from human outbreaks provides neither a biological nor a molecular basis of human H5 and H7 virus pathogenesis. Additionally, criteria for the intravenous pathogenicity index in 6-week-old chickens, which determine the classification of avian influenza viruses as high or low pathogenicity, are independent of the pathogenicity observed in mammals (WHO, 2002). Thus, it is important to utilize mammalian models to elucidate the pathogenicity and possible transmissibility of these viruses. Moreover, the use of both mouse and ferret models provide us with an opportunity to identify and study basic, evolutionarily aspects of influenza virus virulence and host response (Fig. 1). The following sections will discuss the use of mouse and ferret models to study the pathogenesis and transmissibility of avian influenza viruses.

It should be noted that the mouse and ferret, while utilized most frequently, are not the only mammalian models of influenza virus pathogenesis available to researchers. For modeling human influenza infection, primate models offer the advantage of a higher order species, but their use is limited by many practical and ethical constraints (Rimmelzwaan *et al.*, 2001). Additionally, the primate model generally does not exhibit similar clinical symptoms as humans, such as respiratory signs or fever, following infection with many seasonal or avian influenza viruses (Murphy *et al.*, 1982). The guinea pig has emerged as an alternate model for influenza virus transmissibility studies; however, unlike ferrets, guinea pigs do not present with clinical symptoms similar to humans following infection with either human or avian influenza viruses (Lowen

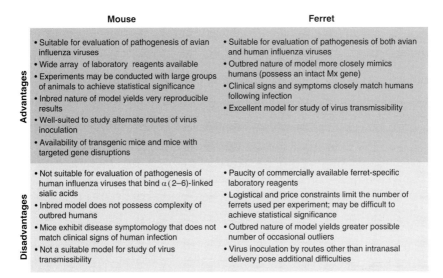

Mouse	Ferret
Advantages • Suitable for evaluation of pathogenesis of avian influenza viruses • Wide array of laboratory reagents available • Experiments may be conducted with large groups of animals to achieve statistical significance • Inbred nature of model yields very reproducible results • Well-suited to study alternate routes of virus inoculation • Availability of transgenic mice and mice with targeted gene disruptions	• Suitable for evaluation of pathogenesis of both avian and human influenza viruses • Outbred nature of model more closely mimics humans (possess an intact Mx gene) • Clinical signs and symptoms closely match humans following infection • Excellent model for study of virus transmissibility
Disadvantages • Not suitable for evaluation of pathogenesis of human influenza viruses that bind α(2–6)-linked sialic acids • Inbred model does not possess complexity of outbred humans • Mice exhibit disease symptomology that does not match clinical signs of human infection • Not a suitable model for study of virus transmissibility	• Paucity of commercially available ferret-specific laboratory reagents • Logistical and price constraints limit the number of ferrets used per experiment; may be difficult to achieve statistical significance • Outbred nature of model yields greater possible number of occasional outliers • Virus inoculation by routes other than intranasal delivery pose additional difficulties

FIGURE 1   Advantages and disadvantages of the mouse and ferret models for use in influenza virus research.

et al., 2006; Van Hoeven *et al.*, 2009a). Nonetheless, the smaller guinea pig species does allow for studying the effect of environmental conditions on virus transmissibility in a confined space. Using this model, Lowen *et al.* recently demonstrated that the transmission of a human strain of influenza virus between guinea pigs is acutely sensitive to both relative humidity and temperature (Lowen *et al.*, 2007, 2008). Further work in this model has also allowed for comparative studies between aerosol and fomite virus transmission (Mubareka *et al.*, 2009). Experimental infection of other mammalian species, including cats and dogs, has additionally furthered our understanding of avian influenza virus pathogenesis (Giese *et al.*, 2008; Rimmelzwaan *et al.*, 2006). Both cats and dogs are susceptible to HPAI H5N1 virus infection, providing additional concern on the zoonotic potential of these species in adaptation of avian influenza viruses and subsequent transmission to humans.

IV. USE OF THE MOUSE MODEL TO STUDY INFLUENZA VIRUS PATHOGENESIS

A. Mouse model for human influenza A virus pathogenesis

The mouse traditionally has been the most common mammalian model used for the study of influenza virus pathogenesis. The relatively low-cost, easy-husbandry, and well-characterized genetics have made this species a favorite for influenza virus research. However, the main disadvantage of this model is that the mouse is not a natural host of this virus, and human influenza A virus subtypes (H1, H2, and H3) generally must be adapted to this species before a virus will replicate efficiently in the murine respiratory tract. This is thought to be due, at least in part, to the paucity of α(2–6) sialic acids (the receptor preferred by human influenza viruses) present in murine respiratory tissues (Ibricevic *et al.*, 2006; Shinya *et al.*, 2006; van Riel *et al.*, 2006). Serial lung passage of human influenza viruses in mice results in the selection of highly virulent variants and this strategy of adaptation has been used to identify specific mutations that may be indicators and predictors of virulence (Brown and Bailly, 1999; Ward, 1997). Although HA changes are likely instrumental in adaptation of human influenza strains to this host, this process has identified amino acid substitutions in multiple virus genes (Brown *et al.*, 2001). Historically, A/Puerto Rico/8/34 (PR/8) and the first human influenza isolate, A/WSN/33 (WSN), have been the most commonly used laboratory mouse-adapted strains. Both H1N1 strains can cause severe lung inflammation depending on the challenge dose and anesthetic used for inoculation. In general, mice infected intranasally with mouse-adapted influenza strains in volumes >20 µL experience lower respiratory tract disease, exhibit a drop in body temperature, and may succumb to virus infection, as compared with humans. Another disadvantage of this model is that infected mice do not shed influenza virus via the respiratory tract (in contrast to humans who shed virus at the upper respiratory mucosal surface) and animals must be euthanized for determination of virus titers in tissue. Despite this difference in disease symptomology, mice serve well as a model for influenza pathogenesis as the onset of symptoms, lung pathology, and cytokine production in mice and humans are temporally related to virus replication (Conn *et al.*, 1995; Hennet *et al.*, 1992; Kurokawa *et al.*, 1996; Peper and Van Campen, 1995; Vacheron *et al.*, 1990; Van Reeth, 2000). When physical signs of illness worsen in BALB/c mice, lethargy, ruffled fur, and weight loss are often used as measurable outcomes and markers for virulence. Fifty percent mouse lethal dose (LD_{50}) titers are the most commonly used lethality indicator. Many measurable cytokines are produced in the infected mouse lung, including IFN-α, IL-1α/β, TNF-α, and IL-6 that can be detected in bronchoalveolar lavage (BAL) fluid and lung homogenates. In general, these cytokines are believed to contribute to the recruitment and activation of nonspecific and virus-specific immune cells (Doherty *et al.*, 1992). Histologic analyses show primarily monocyte/macrophages in the lungs, and the extent of the infiltrate correlates with the virulence of the infecting virus (Wyde and Cate, 1978; Wyde *et al.*, 1978). Interpretation of mouse pathogenesis data obtained from testing mouse-adapted influenza strains requires some consideration of the unique molecular changes that occur as a result of extensive virus passage in laboratory animals. The passage histories of many mouse-adapted strains are unknown and these viruses no longer fairly represent the original progenitor virus. For example, WSN virus, following extensive animal passage since 1933, possesses unique biological properties that are similar to HPAI viruses; the virus replicates in cultured cells without the addition of trypsin and causes systemic infection when higher doses are inoculated intranasally into mice (Castrucci and Kawaoka, 1993). This enhanced virulence factor can be partly attributed to WSN NA plasminogen-binding activity, leading to increased HA cleavage and virus replication in extrapulmonary cells (Goto *et al.*, 2001). Mice have been a staple of influenza virus research for some time. More recently, this model has been a valuable tool for characterizing HPAI viruses as well as dissecting cellular immune responses to influenza infection, and has furthered our understanding of the heightened pathogenicity observed with lethal influenza viruses.

B. Mouse model for H5N1 virus pathogenesis

Many avian influenza H5N1 viruses replicate in mouse lungs to high titers without the prior adaptation that is required for human influenza A viruses. (Gao *et al.*, 1999; Lu *et al.*, 1999; Tumpey *et al.*, 2000). Within this system, the 1997 HPAI H5N1 viruses isolated from humans replicated efficiently in mouse lungs and generally fell into either a high or a low pathogenicity phenotype. The differential pathogenicity phenotypes have been primarily represented by A/Hong Kong/483/97-like (lethal) and A/Hong Kong/486/97-like (nonlethal) viruses. Systemic spread of H5N1 virus, cytokine dysregulation, severe tissue pathology, and death were characteristic of HK/483-like viruses, whereas viruses of the low patho-genicity phenotype, HK/486-like viruses, were limited to replication in the mouse respiratory tract and were usually cleared by days 7–9 postin-fection (p.i.). These H5N1 virus groups also showed pronounced differ-ences in their effects of the innate immune/inflammatory responses that may represent one mechanism of pathogenicity among these lethal viruses in mammalian hosts (Katz *et al.*, 2000a; Tumpey *et al.*, 2000). Similar to the significant lymphopenia among 10 human patients with confirmed H5N1 virus infection (Tran *et al.*, 2004), the number of circulat-ing lymphocytes in mice and ferrets infected with these highly virulent H5N1 viruses was significantly reduced (Maines *et al.*, 2006; Tumpey *et al.*, 2000). Alterations in lymphocyte numbers may be due to the differ-ential induction of apoptosis between highly virulent and low-virulence H5N1 viruses (Tumpey *et al.*, 2000). There was a greater level of apoptosis in HK/483-infected lung and spleen tissues compared with that observed in tissues of HK/486-infected mice. HK/483-infected mice also displayed a dramatic reduction in the CD4–CD8 double-positive thymocyte population and the number of cells harvested from HK/483-infected tissues, such as the thymus, spleen, and lung were significantly lower than the number of tissue cells from HK/486-infected mice that survive infection. Collectively, these data indicate that the highly lethal HK/483 virus targets lymphocytes, resulting in the systemic destruction of these cells in the blood and tissues of infected mice. In addition to finding evidence of lymphocyte destruction, diminished cytokine expression in the lung and spleen tissue was associated with highly virulent H5N1 viruses in mice. Multiple cytokines, such as IFN-γ, IL-1β, and TNF-α, which are typically produced in substantial amounts in the infected lung, were significantly lower in the lungs of mice infected with the highly pathogenic HK/483 virus compared with levels in mice infected with the low pathogenicity HK/486 virus. The well-characterized chemokine MIP-1α, which has been shown to activate and exert chemotactic effects on lymphocytes, macrophages, and neutrophils, had reduced levels in the lungs of HK/483-infected mice (Tumpey *et al.*, 2000). Fewer immune cells migrating into the infected lung tissue may in part explain why HK/483-like viruses are never cleared from the tissue and are so lethal.

Systemic spread of H5N1 virus into the brain tissues also distinguished the lethal HK/483-like from the nonlethal HK/486 viruses. Replication of the highly pathogenic HK/483 virus in the brains of infected mice, along with the induction of IFN-γ, TNF-α, and MIP-1α cytokines in this tissue, correlated with the lethal phenotype of this virus (Park *et al.*, 2002; Tanaka *et al.*, 2003; Tumpey *et al.*, 2000). The local synthesis of proinflammatory cytokines within the brain can lead to anorexia, weight loss, and death, possibly contributing to H5N1 virus pathogenesis in the murine model (Rothwell, 1999). The H5N1 virus A/Hong Kong/156/97, similar to HK/483 virus, was additionally virulent in this model, with high titers of virus in mouse lungs and systemic spread of virus detected before mice succumbed to infection (Gubareva *et al.*, 1998). Following the characterization of the H5N1/97 viruses, the isolation of a HPAI H5N1 virus from duck meat in 2001 represented an unrecognized potential source of human exposure to avian influenza viruses. This virus, A/Dk/Anyang/AVL-1/01 (Dk/Anyang), was of lower pathogenicity for mice as compared with the highly pathogenic 1997 viruses despite possessing a genetically similar HA, but still resulted in lymphopenia, systemic spread of virus, and mortality of up to 50% of infected mice (Lu *et al.*, 2003; Tumpey *et al.*, 2002).

The reemergence of H5N1 viruses in Asia in 2003, and the continued isolation of highly pathogenic viruses within this subtype from humans, has warranted additional study of these viruses. A HPAI H5N1 virus isolated from a human case in 2003, A/Hong Kong/213/03 (HK/213), is unusual in that it can bind to both α(2–3)- and α(2–6)-linked sialic acids

(Shinya *et al.*, 2005). This virus replicated to high titer in the lungs of inoculated mice as well as in extrapulmonary organs, however, was generally not lethal in these mice unless administered at high doses ($>5 \times 10^{5.5}$ EID$_{50}$) (Guan *et al.*, 2004; Shinya *et al.*, 2005). In 2004, multiple Asian countries announced poultry outbreaks due to HPAI H5N1 virus; the outbreaks were widespread in Vietnam and Thailand, with approximately 90% and 60% of provinces affected. The 62% mortality rate among humans with documented H5N1 disease in 2004 and 2005 was markedly higher than the 33% fatality rate among documented human H5N1 cases in 1997. A similar dichotomy of pathogenecity phenotypes in mice were observed among the 2004 H5N1 isolates. Thus, like the HK/483/97 group of viruses, the majority of the 2004 H5N1 isolates exhibited high viral titers in the lung with systemic spread of virus to the thymus, spleen, heart, and brain, substantial weight loss, and lymphopenia in peripheral blood before death of BALB/c mice. Some H5N1 viruses isolated from humans in 2004, including A/Thailand/16/04 (Thai/16) and A/Vietnam/1203/04 (VN/1203), were able to kill mice with as few as 20–60 infectious units in the BALB/c mouse model (Maines *et al.*, 2005). In contrast, replication of H5N1 viruses isolated from avian species and one human isolate, A/Thailand/SP/83/2004 (SP/83), virus was restricted to the respiratory tract and generally resulted in a nonlethal infection (Maines *et al.*, 2005; Muramoto *et al.*, 2006). Furthermore, the lungs of mice infected with Thai/16 virus possess increased cellularity with significantly elevated levels of neutrophils and macrophages throughout the course of infection as compared with SP/83 virus (Perrone *et al.*, 2008). Interestingly, Thai/16 and SP/83 viruses differ by only 13 amino acids, including the Lys/Glu difference at residue 627 (E627K) of PB2 protein. As such, these types of studies have been useful in identifying molecular correlates associated with virulence of HPAI viruses in mammals in order to predict the potential of newly emerging influenza viruses to infect and cause severe disease in humans.

C. Mouse model for H7 virus pathogenesis

As noted above, H7 influenza A viruses were only sporadically associated with human infection until this past decade. Two wild-type H7N7 viruses associated with human conjunctivitis, A/Seal/Mass/1/80 and A/England/268/96, replicated in the respiratory tract of mice but did not cause severe disease in this model (Scheiblauer *et al.*, 1995; T. Tumpey, unpublished data). With increasing reports of H7 human infection, mouse models for avian H7 influenza viruses have recently been established by our laboratory and others to examine the pathogenesis of contemporary viruses within this subtype (Belser *et al.*, 2007a; de Wit *et al.*, 2005; Joseph *et al.*, 2007). The HPAI H7N7 virus A/NL/219/03 (NL/219), isolated from the only fatal human case in The Netherlands outbreak in 2003, was found to be highly lethal in BALB/c mice, as were some equine H7N7 influenza viruses, which were highly lethal for mice without prior adaptation (de Wit *et al.*, 2005; Kawaoka, 1991). NL/219 virus closely resembled H5N1 viruses in this model with respect to their ability to replicate to high titers in the mouse lung, spread systemically and cause cytokine dysregulation (Belser *et al.*, 2007a; de Wit *et al.*, 2005; Joseph *et al.*, 2007). While NL/219 was the only virus from this outbreak to be highly lethal in the mouse, other HPAI H7N7 and H7N1 viruses from the Eurasian lineage replicated to high titer in lungs and were also detected in the brains of infected mice (Belser *et al.*, 2007a; Munster *et al.*, 2007; Rigoni *et al.*, 2007). Replication of HPAI H7N3, LPAI H7N2, and LPAI H7N3 viruses from the North American lineage was generally restricted to the respiratory tract (Belser *et al.*, 2007a; Joseph *et al.*, 2007). Despite the overall reduced virulence of viruses from this lineage in the mouse as compared with Eurasian lineage isolates, North American H7 viruses replicated efficiently in this model without prior adaptation and possessed a low mouse infectious dose (MID_{50}), indicating that viruses within this subtype are capable of high infectivity *in vivo* in the absence of pronounced morbidity or mortality (Belser *et al.*, 2007a).

D. Gene knockout mice in the study of avian influenza

It has been hypothesized that virulent strains of influenza cause more severe pathology in the respiratory tract of infected mammals resulting in the induction of either dysregulated or exacerbated cytokine profiles in the lung, and consequently differences in clinical symptoms during influenza A virus disease (Van Reeth, 2000). Excessive immune cell infiltration during an acute lung injury may impair tissue restoration directly by interfering with gas exchange, or indirectly through the release of soluble immune mediators. Knockout mice deficient in immune mediators allow for a unique opportunity to better understand the contribution of individual host responses in the overall pathogenesis of avian influenza viruses. Both H5N1 viruses isolated from 1997 and 2004 exhibited similar weight loss and lethal disease in B6/129 mice, the mouse strain of most commonly used cytokine- or chemokine-deficient mice, as what is observed in the BALB/c model (Salomon *et al.*, 2007a; Szretter *et al.*, 2007). Mice deficient in IL-6, MIP-1α, or CC chemokine ligand 2 (CCL2) exhibited similar kinetics of weight loss and mortality following infection with HPAI H5N1 viruses as compared with wild-type controls (Salomon *et al.*, 2007a; Szretter *et al.*, 2007). Although H5N1 virus infection elicited strong IL-6, MIP-1α, or CCL2 production in the mouse lung, the data suggest that these proteins are not significantly contributing to the pathogenesis of H5N1 virus or protection from acute infection. However, for single gene knockout studies there is always the concern that other cytokines with overlapping functions can compensate for the loss of a single protein. In support of elevated proinflammatory cytokine response ("cytokine storm") contributing to the pathogenesis of H5N1 virus infections, tumor necrosis factor alpha (TNF-α) has attracted the greatest attention because it is a key regulator of inflammation and H5N1 virus has been shown to be a potent inducer of this proinflammatory cytokine in human primary macrophages (Cheung *et al.*, 2002). TNF-α may contribute to early disease severity, as the absence of TNFR1 signaling significantly delayed morbidity as compared with control mice (Szretter *et al.*, 2007). However, TNFR1-deficient mice displayed similar virus replication and disease outcomes as the wild-type control mice (Salomon *et al.*, 2007a; Szretter *et al.*, 2007). With respect to cytokines providing a protective role, it was found that infection of IL-1R-deficient mice with HK/486 virus resulted in heightened morbidity and mortality compared with control mice (Szretter *et al.*, 2007). In addition, HK/486-infected mice deficient in type 1 interferon (IFN-αβR) demonstrated higher viral titers and increased time-to-death as compared with infected control mice (Szretter *et al.*, 2009). To further explore the role of hypercytokinemia in H5N1 infection, NF-κB p50 knockout mice were infected with the H5N1 virus A/mallard/Bavaria/1/06 (Droebner *et al.*, 2008). While the lungs of wild-type mice displayed elevated levels of cytokines and chemokines, NF-κB p50 knockout mice displayed a strong reduction in these factors following infection. However, no differences in the lethality, viral titers, or systemic spread of virus were observed between wild-type and knockout mice. Initial studies with cytokine-deficient mice have demonstrated an important role for selected host proinflammatory cytokines in the mitigation of influenza virus infection, but should be taken into consideration overall as these data would suggest that new anticytokine storm therapeutics may not universally improve disease progression.

In addition to host cytokine responses, transgenic mice have allowed for the study of other host factors, such as Mx proteins, which are a family of GTPases that are specifically associated with conferring resistance to orthomyxoviruses (Haller *et al.*, 2006; Staeheli *et al.*, 1993). The *Mx1* gene is under tight transcriptional control of alpha/beta interferon (IFN-α/β) and codes for a nuclear 72-kDa protein. Standard laboratory BALB/c mice carry defective alleles of the *Mx1* gene and although there are differences between the Mx systems of humans and mice, $Mx1^{+/+}$ mice which carry functional *Mx1* alleles may better mimic the innate immune system of humans. Exploring the role of this protein on host survival, we and others found that $Mx1^{+/+}$ mice survive infection with VN/1203 virus and exhibit reduced viral replication in the lung and brain as compared with standard laboratory $Mx1^{-/-}$ mice (Salomon *et al.*, 2007b; Tumpey *et al.*, 2007b). Moreover, treatment of $Mx1^{+/+}$ mice with recombinant human IFN-α increased the resistance to the H5N1 virus not observed in similarly treated standard BALB/c mice. The importance of interferon response products like Mx1 in the clearance of influenza virus infection have been demonstrated and polymorphisms in these genes and others found within the human population could conceivably contribute to the spectrum of virulence observed with influenza virus infected patients (Dupuis *et al.*, 2003; Nakajima *et al.*, 2007). These studies clearly demonstrate the importance of the host's interferon response in controlling avian influenza viruses.

E. Tropism of avian influenza viruses

The versatility of the mouse model allows for the examination of additional properties that may contribute to influenza virus pathogenesis, such as the ability of viruses to infect by alternate routes of inoculation. The question of why some H5N1 isolates spread systemically, whereas others do not, was addressed by testing nontraditional routes of inoculation. Following intravenous or intracranial inoculation, it was found that HK/483 virus replicated in the lung and brain of mice, with all mice succumbing to infection by day 9 p.i., further demonstrating the high virulence of this virus in the mouse previously observed following intranasal inoculation (Bright *et al.*, 2003). Intracranial inoculation with the nonlethal HK/486 virus also resulted in a fatal infection; however, virus replication was limited to the brain. Moreover, HK/486 virus did not replicate in the lung or brain following intravenous inoculation (Bright *et al.*, 2003). Additional studies are needed to determine the precise virulence determinants in this model, but genetic studies would suggest that lysine at position 627 of the PB2 protein confers efficient HK/483 virus replication in extrapulmonary tissues (Shinya *et al.*, 2004).

Conjunctivitis is frequently reported following human infection with H7 influenza viruses, and infrequently observed following infection with other avian and human virus subtypes (Olofsson *et al.*, 2005). As such, our laboratory established a model of ocular inoculation in the BALB/c mouse model to study the ability of influenza viruses of multiple subtypes to use this tissue as a portal of entry (Belser *et al.*, 2007a). Use of this model revealed the ability of Eurasian lineage H7N7 and North American lineage H7N3 HPAI viruses to replicate in the mouse eye following ocular inoculation. Moreover, the H7N7 virus (NL/219) spread to the lung tissue and resulted in a fatal infection among 30% of inoculated mice. HPAI H5N1 viruses replicated to overall lower titers in the eye as compared with H7 viruses, and were most frequently detected in the nose and lung following ocular inoculation (Belser *et al.*, 2009). Two HPAI H5N1 viruses, HK/483 and Thai/16, were capable of mounting a lethal infection in 60%

of mice following ocular inoculation. Conversely, human H3N2 and H1N1 viruses did not replicate in the eye following ocular inoculation and were only sporadically detected in the lung postinfection (Belser *et al.*, 2007a, 2009; Tannock *et al.*, 1985). These studies demonstrate the ability of avian influenza viruses to mount productive and lethal infections following ocular inoculation. This finding underscores the importance of wearing personal protective equipment, including eye protection which has been recommended for persons involved in avian influenza outbreak responses, in the event of possible exposure to avian influenza viruses (CDC, 2006).

V. USE OF THE FERRET MODEL TO STUDY INFLUENZA VIRUS PATHOGENESIS

A. Ferret model for human influenza A virus pathogenesis

Unlike the mouse, human influenza A viruses infect the ferret without the requirement for prior host adaptation. Ferrets are well-suited for the study of virus pathogenesis as this species exhibits clinical symptoms following influenza virus infection, such as sneezing, fever, and nasal discharge, that closely models influenza infection of humans (Smith and Sweet, 1988; Sweet *et al.*, 1979). There are two main lines of evidence indicating that the respiratory tract of ferrets closely resembles that of humans; first, ferrets possess a predominance of $\alpha(2\text{--}6)$-linked sialic acids on the upper airway epithelia (Ibricevic *et al.*, 2006; Leigh *et al.*, 1995; Maher and DeStefano, 2004), and second, avian H5N1 and human H3N2 influenza viruses exhibit similar patterns of virus attachment to tissues from both species (van Riel *et al.*, 2006, 2007). Replication of human influenza viruses of low virulence in ferrets is generally restricted to the respiratory tract (Basarab and Smith, 1969; Cavanagh *et al.*, 1979; Haff *et al.*, 1966); however, human H3N2 isolates have been isolated from brain tissue among ferrets that did not exhibit any severe clinical signs of disease (Zitzow *et al.*, 2002). This could be due to the proximity of high titers of infectious virus detected in nasal turbinates following intranasal inoculation, with subsequent spread to the olfactory bulb and brain. Thus, influenza A virus found in the ferret brain tissue may not be an indicator of extrapulmonary spread and the level of virulence, as most low-virulent strains cannot be detected in other systemic tissues of the ferret (Zitzow *et al.*, 2002). This is in contrast to the efficient spread of HPAI H5N1 virus to multiple extrapulmonary organs, including the spleen, intestine, liver, and peripheral blood of ferrets (Maines *et al.*, 2005).

B. Ferret model for H5N1 virus pathogenesis

The ferret has become an important model for the study of avian influenza virus pathogenesis. Avian influenza viruses of low pathogenicity have been studied in the ferret, with most resulting in mild infection accompanied with efficient virus replication in the upper and lower respiratory tract as well as the intestinal tract postinoculation (Hinshaw *et al.*, 1981; Kawaoka *et al.*, 1987). For HPAI viruses, the dichotomy of pathogenicity phenotypes among HK/483 and HK/486 virus-infected mice (Lu *et al.*, 1999) was not observed in ferrets. Both HK/483 and HK/486 H5N1 viruses caused severe disease in ferrets, characterized by lethargy, clinical signs of respiratory disease, weight loss, transient lymphopenia, and neurological signs in some animals (Zitzow *et al.*, 2002). Following intranasal inoculation, virus replicated to high titer in nasal washes and nasal turbinates with titers $>10^4$ EID$_{50}$ sustained through day 5 p.i. Systemic spread of virus was observed following inoculation with both H5N1/97 viruses, with virus recovered from the lung, brain, spleen, and intestine. These results suggest that a lysine at position 627 of PB2 protein is not necessary for a virulent phenotype in this model. Other substitutions in PB2 or within other H5N1 virus genes likely contribute to the high pathogenicity phenotype observed in ferrets. Not all HPAI H5N1 viruses are virulent in ferrets; Dk/Anyang (2001 isolate) was asymptomatic in this model, with virus recovered from nasal washes but not the lungs or extrapulmonary tissues following inoculation (Lu *et al.*, 2003). Ferret pathotyping of the 2003 H5N1 virus, HK/213, yielded different results among different laboratories. In most studies, HK/213 virus replicated in the nasal turbinates and lungs of ferrets but generally resulted in mild infection; however, in one study HK/213 virus caused lethal disease including lower respiratory tract virus replication, substantial weight loss, and hind-limb paralysis (Maines *et al.*, 2006; Shinya *et al.*, 2005; Webby *et al.*, 2004; Yen *et al.*, 2007). This may be due to multiple variations in experimental conditions between studies, including the age of the ferrets, the inoculum viral dose, the number of nasal washings, and the anesthetic dose per animal.

The HPAI H5N1 viruses isolated from 2004 exhibited enhanced virulence in the ferret model compared with previous isolates (Govorkova *et al.*, 2005; Maines *et al.*, 2005). Viruses that exhibited a high pathogenicity phenotype in ferrets caused severe clinical signs of illness, including lethargy, severe weight loss, lymphopenia, and neurological symptoms in some animals. Systemic spread of virus to extrapulmonary organs, including the brain, spleen, and intestine, was frequently observed. These viruses, including VN/1203 and Thai/16, resulted in a more rapid mean time-to-death of inoculated ferrets as compared with HPAI viruses isolated from 1997 (Maines *et al.*, 2005). Histopathologic evaluation revealed diffuse interstitial inflammation in the lungs as well as inflammation in the brains of ferrets inoculated with viruses of high virulence as compared with viruses of low virulence (Govorkova *et al.*, 2005; Maines *et al.*, 2005). Similar severe disease was observed with A/ Indonesia/5/05 (Indo/05) and A/Vietnam/JP36-2/05 viruses (Maines *et al.*, 2006; Yen *et al.*, 2007). However, a more recent H5N1 virus, A/ Turkey/15/06, exhibited reduced virulence as compared with VN/1203 virus in the ferret and was generally not lethal unless administered at a high dose of virus (10^7 EID_{50}) (Govorkova *et al.*, 2007). Inoculation of ferrets with A/Turkey/65-596/06 virus additionally resulted in mild infection, with virus recovered from nasal washes and respiratory tissues but not the brain (Yen *et al.*, 2007). It was recently found that either consumption or intragastric administration of VN/1203 virus-infected meat resulted in a lethal infection in ferrets, demonstrating the ability of H5N1 virus to initiate infection through the digestive system (Lipatov *et al.*, 2009). The use of the ferret model has allowed for greater examination of avian influenza H5N1 virus pathogenesis in mammalian hosts, allowing for study of clinical parameters following infection that cannot be as closely studied in other animal models such as the mouse.

C. Ferret model for H7 virus pathogenesis

As with H5N1 viruses, the ferret model has furthered our understanding of the capacity of H7 avian influenza viruses to cause disease. Contemporary Eurasian lineage HPAI H7N7 viruses (NL/219 and A/NL/230/03; NL/230) demonstrated enhanced virulence in the ferret model as compared with North American lineage H7N2 viruses (Belser *et al.*, 2007a). Both NL/219 and NL/230 viruses replicated to high titers in the upper and lower respiratory tract of ferrets with virus isolated from the brain, intestine, and other systemic organs following virus inoculation. NL/219 virus exhibited high virulence in this model; infected ferrets exhibiting lethargy, severe weight loss, and lymphopenia, with neurological symptoms in some animals. In contrast, infection with LPAI H7N2 North American viruses was generally mild, exhibiting no lethargy and only modest weight loss. Virus replicated efficiently in the nasal washes of inoculated ferrets and was detected at high titer in nasal turbinates but at reduced titers in the ferret lung and extrapulmonary tissues (Belser *et al.*, 2007a). Inoculation with a HPAI H7N3 virus isolated from Canada in 2004 resulted in increased morbidity in the ferret model compared with other North American isolates but did not result in lethal disease (Belser *et al.*, 2008). These studies have revealed that HPAI viruses of multiple subtypes have the capacity to cause severe disease and death in ferrets, and demonstrated that avian influenza viruses that pose a pandemic threat are not limited to those within the H5 subtype.

D. Transmissibility of avian influenza viruses

Influenza virus is a highly contagious respiratory pathogen and can spread rapidly between susceptible individuals during epidemics or pandemics. Person-to-person transmission of influenza virus can occur by direct or indirect contact, or by respiratory droplets that are expelled during coughing or sneezing (Bridges *et al.*, 2003). However, the molecular determinants that govern virus transmissibility and the optimal route of transmission for efficient spread of virus within a community are currently not fully understood. To address these questions, our laboratory and others have utilized the ferret model to study the transmissibility of human and avian influenza viruses. Due to the high diversity of influenza viruses, it is important to assess both potential routes—that is, transmission occurring due to direct or indirect contact and/or transmission occurring only by respiratory droplets—to best understand the transmissibility of a given virus. To demonstrate that human H3N2 viruses transmit via direct contact, inoculated and contact (naïve) ferrets were housed in the same cage, sharing food, drink, and bedding (Herlocher *et al.*, 2001; Yen *et al.*, 2005a). To investigate the ability of viruses to transmit by respiratory droplets (droplet or droplet nuclei), our laboratory developed a model that allowed for air exchange while preventing direct contact of inoculated and contact ferrets. To achieve this, ferrets were housed in adjacent cages, each with a perforated side wall (Maines *et al.*, 2006). These models have demonstrated the ability of human H3N2 and H1N1 viruses to transmit efficiently between ferrets by respiratory droplets, as measured by detection of virus titers in nasal washes and seroconversion for hemagglutination inhibition (HI) antibody in all contact animals (Maines *et al.*, 2006; Tumpey *et al.*, 2007a).

Despite the high virulence of H5N1 viruses in humans, human-to-human transmission of H5N1 viruses has been only rarely documented (Kandun *et al.*, 2006; Olsen *et al.*, 2005; Ungchusak *et al.*, 2005). However, the increased detection of HPAI H5N1 viruses in wild birds and poultry, and the escalating number of confirmed human cases, emphasize the need to better understand the capacity of H5N1 viruses to transmit between mammals and the risk for reassortant avian–human viruses to acquire this property. These studies have found that H5N1 viruses, isolated from the original 1997 outbreak or from 2003 to 2005, do not transmit efficiently by either respiratory droplets or direct contact in ferrets (Maines *et al.*, 2006; Yen *et al.*, 2007). Inefficient transmission of the H5N1 viruses HK/486 and A/Vietnam/JP36-2/05 was observed as seroconversion was detected in contact ferrets, with virus in nasal wash and severe disease sporadically detected postcontact. In contrast, other H5N1 viruses tested, including HK/213 and Indo/05 viruses, did not transmit by either direct contact or respiratory droplets in ferrets (Maines *et al.*, 2006; Yen *et al.*, 2007). To determine if reassortment with a human virus could enhance the transmissibility of an avian virus, reassortant viruses containing various gene segments from the avian H5N1 virus HK/486 and the human H3N2 virus A/Victoria/3/75 were generated (Maines *et al.*, 2006). This work revealed that human viruses bearing avian HA and NA surface glycoproteins (or the reciprocal constellation) did not retain the efficient transmission by respiratory droplets observed with the parental H3N2 virus. Additionally, the substitution of the human virus ribonucleoprotein genes (PB2, PB1, PA, NP) did not enhance the transmissibility of the avian virus (Maines *et al.*, 2006), suggesting that additional human influenza virus genes are required for efficient transmission in mammals. Studies utilizing the reconstructed 1918 virus have identified that a reassortant virus bearing the HA and PB2 genes from this pandemic strain, with all remaining genes derived from an avian H1N1 virus, is sufficient to confer respiratory droplet transmissibility in the ferret model; further work will reveal if the same is true for avian viruses within other subtypes (Van Hoeven *et al.*, 2009b).

As limited human-to-human transmission of H7 viruses has been reported (Koopmans *et al.*, 2004), viruses within this subtype were additionally tested for their ability to transmit by either direct contact or respiratory droplets (Belser *et al.*, 2008). One LPAI H7N2 virus, A/NY/107/03 (NY/107), isolated from an individual with respiratory symptoms (CDC, 2004b), transmitted efficiently by direct contact, with virus detected in nasal washes of contact ferrets as early as day 2 postcontact. Transmission was not observed with other contemporary North American lineage H7N2 or H7N3 viruses tested. Similar to H5N1 viruses, the HPAI H7N7 NL/219 virus was highly virulent in the ferret model but not transmissible between ferrets. However, NL/230 virus displayed enhanced transmissibility in this model by direct contact as compared with NL/219 virus, with virus isolated from nasal washes and seroconversion for HI antibody detected in two of three contact ferrets. In comparison with selected H7 viruses that demonstrated the ability to transmit by direct contact, none of the H7 viruses tested transmitted by respiratory droplets (Belser *et al.*, 2008). Demonstrating the need for continued surveillance and study of influenza viruses, both an H2N3 human/avian/swine triple reassortant virus isolated from swine in the United States in 2006, as well as selected avian H9N2 viruses, were recently shown to transmit efficiently in the ferret model by direct contact (Ma *et al.*, 2007; Wan *et al.*, 2008).

VI. MOLECULAR BASIS OF AVIAN INFLUENZA PATHOGENESIS

The use of animal models to assess the pathogenesis of avian influenza A viruses has readily demonstrated that not all HPAI viruses are highly virulent in mammals. Additionally, the capacity of a virus to mount a productive and lethal infection can vary between species, demonstrating that influenza virulence is determined by both viral determinants and host factors. The use of plasmid-based reverse genetics techniques has allowed for a more detailed study of the contribution of individual gene segments on virus pathogenicity (Fodor *et al.*, 1999; Neumann and Kawaoka, 2002; Neumann *et al.*, 1999). These studies have revealed crucial roles of individual viral proteins and often single amino acid positions that dramatically affect the virulence of selected avian influenza viruses. However, despite these individual contributions to virulence ascribed to certain viral proteins which are described in more detail below, it is apparent that the overall composition of viral gene products and host determinates, rather than any one particular mutation, is responsible for virulence (Fig. 2) (Katz *et al.*, 2000b).

A. Hemagglutinin cleavage site

Viruses of the H5 and H7 subtype can acquire molecular features in the HA cleavage site that result in enhanced pathogenicity for land-based poultry. The posttranslational cleavage at a conserved arginine residue of HA0 into the subunits HA1 and HA2 is necessary for virus infectivity as it activates the membrane fusion potential of the HA (Skehel and Wiley, 2000). The presence of a single arginine residue is sufficient for cleavage to occur, and in general human influenza viruses contain only one arginine at the cleavage site; human viruses are cleaved by extracellular trypsin-like

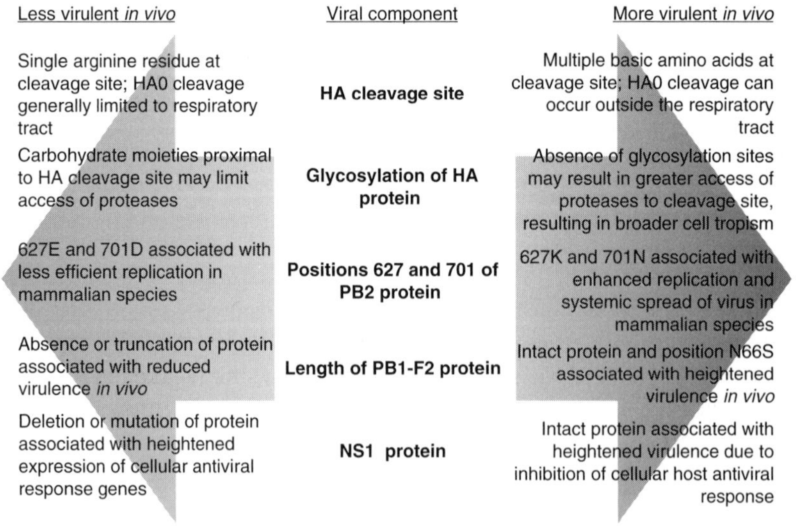

Less virulent *in vivo*	Viral component	More virulent *in vivo*
Single arginine residue at cleavage site; HA0 cleavage generally limited to respiratory tract	**HA cleavage site**	Multiple basic amino acids at cleavage site; HA0 cleavage can occur outside the respiratory tract
Carbohydrate moieties proximal to HA cleavage site may limit access of proteases	**Glycosylation of HA protein**	Absence of glycosylation sites may result in greater access of proteases to cleavage site, resulting in broader cell tropism
627E and 701D associated with less efficient replication in mammalian species	**Positions 627 and 701 of PB2 protein**	627K and 701N associated with enhanced replication and systemic spread of virus in mammalian species
Absence or truncation of protein associated with reduced virulence *in vivo*	**Length of PB1-F2 protein**	Intact protein and position N66S associated with heightened virulence *in vivo*
Deletion or mutation of protein associated with heightened expression of cellular antiviral response genes	**NS1 protein**	Intact protein associated with heightened virulence due to inhibition of cellular host antiviral response

FIGURE 2   Selected molecular determinants of avian influenza virus pathogenesis.

proteases that are generally limited to the respiratory tract (Bottcher *et al.*, 2006). Likewise, the majority of avian HA subtypes possess only a single arginine and are cleaved by proteases present in the intestinal epithelium (Steinhauer, 1999). However, HPAI viruses of the H5 and H7 subtype contain multiple basic amino acids at the cleavage site, which allow the virus to be cleaved by ubiquitously expressed intracellular proteases, including furin-like proteases (Bosch *et al.*, 1981; Steinhauer, 1999; Walker *et al.*, 1994). These highly pathogenic viruses can arise from less virulent strains by insertion of multiple basic amino acids at the cleavage site due to the polymerase stalling during transcription, resulting in the stepwise substitution of amino acids at the cleavage site, during adaptation to land-based poultry (Perdue *et al.*, 1997; Steinhauer, 1999). In mammalian hosts, the multibasic amino acid insertion within the cleavage loop of the H5N1 HA protein is necessary for a highly virulent phenotype as removal of this sequence or replacement of this sequence with that from avirulent avian strains results in virus attenuation in mice (Hatta *et al.*, 2001). However, possession of a multibasic amino acid cleavage site is not sufficient for virulence, as not all HPAI H5 or H7 viruses which contain this feature are lethal in mammals (Belser *et al.*, 2007a; Maines *et al.*, 2005; Tumpey *et al.*, 2000). Unlike LPAI H5 viruses, which can acquire a high pathogenicity phenotype by mutation or insertion of basic amino acids at the cleavage site, it is believed that LPAI H7 viruses require an insertional event to achieve this level of virulence (Horimoto *et al.*, 1995; Lee *et al.*, 2006; Webster *et al.*, 1986). In the case of two H7 outbreaks, viruses of high pathogenicity emerged following non-homologous recombination events, resulting in the insertion of multiple amino acids derived from other viral genes into the cleavage site (Hirst *et al.*, 2004; Suarez *et al.*, 2004). Additionally, the insertion of three arginine residues at the HA cleavage site of A/Seal/Mass/1/80, an H7N7 virus of low pathogenicity associated with a case of human conjunctivitis, resulted in a virus with enhanced pathogenicity in mice and ferrets (Scheiblauer *et al.*, 1995; Webster *et al.*, 1981).

B. Surface glycoproteins (HA and NA)

An evolutionary balance exists between the HA and NA surface proteins of influenza viruses (Mitnaul *et al.*, 2000). There is a correlation between the avidity of HA and the strength of NA activity; the weaker the receptor binding of the HA, the weaker the activity of the NA (Wagner *et al.*, 2000, 2002). Mutations within the active sites of these proteins moderate their binding activities, while insertions and deletions in the stalk regions of the NA that affect the length of the protein also modify the enzymatic activity of NA (Baigent *et al.*, 1999; Baigent and McCauley, 2001). Over the last decade, NA stalk deletions have become more prevalent in H5N1 viruses isolated from avian species; recent work has found that H5N1 viruses with short NA stalks are more virulent in mice compared with viruses possessing long NA stalks (Matsuoka *et al.*, 2009). Stalk deletions in the NA that removed potential glycosylation sites, in addition to acquisition of glycosylation sites in the HA receptor-binding site, were observed among HPAI H7N1 viruses, unlike many LPAI viruses tested (Banks *et al.*, 2001). Glycosylation of HA of avian influenza A viruses may decrease the dependence on NA for virus release from host cells (Baigent and McCauley, 2001; Hulse *et al.*, 2004; Ohuchi *et al.*, 1995, 1997). Furthermore, the absence of carbohydrate moieties in the HA stalk region may increase the virulence of a virus by allowing greater access of proteases to the cleavage site (Deshpande *et al.*, 1987). Minor changes in the receptor-binding domain have also been shown to alter the virulence of H5N1 viruses in mice (Yen *et al.*, 2009). The presence of a carboxyl-terminal lysine that conferred the ability to bind plasminogen as well as the absence of a glycoslyation site in the NA was shown to increase the efficiency of HA0 cleavage and broaden cell tropism of the human virus A/WSN/33 (Goto and Kawaoka, 1998; Goto *et al.*, 2001; Li *et al.*, 1993). This also appears to hold true for avian influenza viruses, as a T223I substitution in the NA which eliminated a possible glycosylation site correlated with H5N1 viruses isolated from humans that exhibited a high pathogenicity phenotype in mice (Katz *et al.*, 2000b).

Examination of the contribution of surface glycoproteins to the high virulence of selected H5N1 viruses has revealed that the HA and NA by themselves are not the sole determinants of virus pathogenicity. Exchange of the HA and NA from HK/486 virus with the surface glycoproteins of HK/483 virus resulted in a reassortant virus 100-fold more virulent compared with HK/486 virus (Chen *et al.*, 2007). However, reassortant viruses that possess the surface glycoproteins from a virulent HPAI H5N1 virus with all internal genes derived from an avian virus or human virus that was not lethal in mammalian models did not result in increased pathogenicity in mice or ferrets (Maines *et al.*, 2006; Salomon *et al.*, 2006). Similarly, reassortant viruses possessing either the HA or NA from the HPAI H7N7 NL/219 virus with remaining genes derived from A/NL/33/03 (NL/33, a HPAI H7N7 virus that is not highly virulent in mice) did not exhibit the highly lethal phenotype observed with wild-type NL/219 virus in mice (Munster *et al.*, 2007).

C. Polymerase complex

Numerous studies have demonstrated that the internal genes of influenza virus play a role in virulence (Rott *et al.*, 1979; Snyder *et al.*, 1987). The polymerase proteins PB2, PB1, and PA, along with the nucleocapsid protein (NP), form the ribonucleoprotein complex (RNP) of influenza

A viruses (Lamb and Choppin, 1976). The polymerase proteins form a complex that possesses RNA-dependent RNA polymerase activity and has nuclear localization signals to facilitate transcription of viral (v)RNA while utilizing host cell machinery (Akkina *et al.*, 1987). To determine the contribution of the polymerase complex in the virulence of avian influenza viruses, reassortant viruses containing gene segments from two HPAI H5N1 viruses with differing virulence in mice and ferrets were generated (Salomon *et al.*, 2006). This work revealed that reassortant viruses bearing the polymerase complex (PB2, PB1, PA) from the virus A/Ck/Vietnam/C58/04 (VN/C58) (HPAI H5N1 virus that is not highly virulent in mammalian models) with all remaining genes derived from the highly virulent VN/1203 virus abolished the high pathogenicity observed with the wild-type VN/1203 virus in both mice and ferrets. Accordingly, the reciprocal constellation of genes with the polymerase complex derived from VN/1203 virus, with all remaining genes derived from the chicken virus, was sufficient to recapitulate the highly virulent phenotype in mice, but not ferrets (Salomon *et al.*, 2006). The high virulence of HK/486 virus in ferrets was similarly abolished with a reassortant virus that contained the ribonucleoprotein complex genes (PB2, PB1, PA, NP) from a human H3N2 virus with remaining genes derived from the H5N1 virus (Maines *et al.*, 2006). These studies reveal that changes limited to the polymerase genes are sufficient to substantially alter virus pathogenicity. A recent study found that incorporation of the avian PB1 gene from the Thai/16 virus into the background of the human H3N2 virus A/Wyoming/3/03 resulted in a significant increase in virulence in the mouse (Chen *et al.*, 2008). All three pandemic viruses from the twentieth century possessed an avian PB1 gene, further demonstrating the importance of studying the contribution of polymerase genes to virulence and transmissibility of viruses with pandemic potential (Kawaoka *et al.*, 1989; Taubenberger *et al.*, 2005)

D. PB2 protein

The influenza PB2 protein is involved in the recognition and cleavage of m^7GpppX^m-containing cap structures of host mRNAs which are subsequently used for viral mRNA synthesis (Almond, 1977; Nakagawa *et al.*, 1995; Plotch *et al.*, 1981). An amino acid substitution in the PB2 protein, E627K, has been associated with adaptation and virulence of some influenza A viruses in mice (Hatta *et al.*, 2001; Munster *et al.*, 2007; Rigoni *et al.*, 2007; Subbarao *et al.*, 1993). HPAI H5N1 viruses possessing a lysine at position 627 have been shown to replicate more efficiently in the lungs and nasal turbinates of mice compared with viruses with glutamate at this position (Hatta *et al.*, 2007). Conversely, substituting lysine for glutamate attenuated the highly virulent phenotype observed following infection with HPAI H5N1 and H7N7 viruses in mice and ferrets (Munster *et al.*, 2007; Salomon *et al.*, 2006). The presence of lysine at position 627 results in avian viruses acquiring an enhanced ability to replicate *in vitro* at 33 °C, the temperature of the human upper respiratory tract (Massin *et al.*, 2001). This mutation has also been associated with increased efficiency of virus replication and the ability to outpace the host's immune system in mice (Shinya *et al.*, 2004). A lysine at this position may also contribute to optimal RNA conformational changes during transcription/replication in mammalian cells, and is one of numerous amino acids within the PB2 protein thought to play a role in determination of host range (Crescenzo-Chaigne *et al.*, 2002; Yao *et al.*, 2001).

Studies have revealed that not all highly pathogenic viruses must possess the E627K substitution to be highly pathogenic in mammalian models (Govorkova *et al.*, 2005; Maines *et al.*, 2005; Zitzow *et al.*, 2002). Additionally, the pathogenicity of viruses containing this mutation has been shown to vary depending on the mammalian model. With regard to the two prototypical 1997 H5N1 influenza viruses, HK/483 and HK/486, as well as HPAI H5N1 viruses isolated from 2004, this single E627K amino acid substitution correlated with the heightened pathogenicity phenotype of viruses containing the mutation in mice but not in ferrets (Hatta *et al.*, 2001; Maines *et al.*, 2005; Zitzow *et al.*, 2002). These studies have indicated that a lysine at position 627 of PB2 is not necessary for a virulent phenotype in these models. In support of this, no clear correlation was observed between the amino acid at position 627 of PB2 and the clinical outcome of H5N1 human infections during 2004–2005 (de Jong *et al.*, 2006). Therefore, other substitutions in PB2 or within other genes are also likely to contribute to the high pathogenicity phenotype, consistent with the concept that influenza virus virulence is a polygenic trait (Chen *et al.*, 2007). For example, viruses of multiple subtypes possessing an amino acid substitution at position 701 of PB2 have been associated with systemic spread of virus and enhanced mortality in the mouse model (Gabriel *et al.*, 2005; Li *et al.*, 2005). Following inoculation in mice, H5N1 virus that contained asparagine at position 701 exhibited enhanced replication, systemic spread, and heightened virulence as compared with a virus with a N701D point mutation that possessed aspartic acid at this position (Li *et al.*, 2005). The D701N mutation in PB2 was also associated with enhanced polymerase activity and pathogenicity in the mouse following infection with an H7N7 virus (Gabriel *et al.*, 2005). Recent work has additionally revealed that positions 627 and 701 influenza the transmissibility of influenza viruses in the guinea pig model (Steel *et al.*, 2009). It is clear from these studies that mutations within the PB2 protein can dramatically alter the virulence of avian influenza viruses of multiple subtypes. The identification of single amino acid substitutions within virus proteins that are associated with altered pathogenicity in mammalian models, as demonstrated above, is a valuable tool to more accurately predict the virus pathogenicity in mammalian models.

E. PB1-F2 protein

The PB1-F2 protein is produced by an alternate reading frame within the PB1 gene and is present in varying lengths in all influenza viruses subtypes (Chen *et al.*, 2001; Zell *et al.*, 2007). This proapoptotic protein localizes to the inner and outer membrane of host mitochondria, and can specifically target and destroy alveolar macrophages (Chen *et al.*, 2001; Coleman, 2007; Zamarin *et al.*, 2005). Compared with viruses with an intact PB1-F2, influenza virus PB1-F2 knockout mutants were less virulent in mice, suggesting that this protein contributes to viral pathogenicity *in vivo*, possibly by delaying virus clearance from the lungs (Zamarin *et al.*, 2006). PB1-F2 expression has also been shown to contribute to inflammation following influenza virus infection in mice as well as increase the susceptibility to secondary bacterial pneumonia (McAuley *et al.*, 2007). An amino acid substitution in the PB1-F2 protein, N66S, was recently identified as a potential virulence marker, as it was observed that all HPAI H5N1 viruses isolated from 1997 possessing this substitution exhibited a high pathogenicity phenotype in mice (Conenello *et al.*, 2007). To determine the effect of this mutation on virulence, chimeric viruses that contained the HPAI H5N1 virus A/Hong Kong/156/97 (HK/156) PB1 gene on an A/WSN/33 background with either an asparagine (found in wild-type HK/156 virus) or a serine at amino acid 66 were constructed. Mice infected with the N66S virus displayed increased weight loss, higher lung titers, and delayed viral clearance as compared with the virus that contained asparagine at this position (Conenello *et al.*, 2007). Future work will allow for a greater understanding of this protein and its role during avian influenza A virus infection.

F. NS1 protein

The nonstructural influenza protein NS1 functions as an antagonist to block the IFN-α/β-mediated host antiviral response following infection, and as such has been proposed to be a determinant of influenza virus virulence (Garcia-Sastre, 2001). NS1 has a nuclear localization signal, an RNA-binding region and an effector domain that interferes with host cell machinery (Chen and Krug, 2000; Chen *et al.*, 1999; Fortes *et al.*, 1994; Nemeroff *et al.*, 1998). The antiviral activity of this protein includes sequestering dsRNA generated during virus replication, inhibiting PKR activity by binding to free double-stranded RNA, and preventing transcription of antiviral genes (Garcia-Sastre, 2001; Ludwig *et al.*, 1999; Tan and Katze, 1998; Wang *et al.*, 2000). Deletion of NS1 results in heightened expression of cellular genes involved in the antiviral response, including retinoic acid inducible gene I (RIG-I); NS1 protein has been shown to inhibit RIG-I-induced signaling (Geiss *et al.*, 2002; Guo *et al.*, 2007; Opitz *et al.*, 2007). The use of animal models has furthered our understanding of the contribution of NS1 to virus pathogenesis. A single-gene reassortant virus that combined the NS gene of VN/C58 virus with the remaining seven genes from the highly virulent VN/1203 virus attenuated the virulence observed with the parental VN/1203 virus in ferrets, but not in mice (Salomon *et al.*, 2006). However, a reassortant virus that possessed the NS gene from the highly virulent H7N7 virus NL/219 with remaining genes derived from A/NL/33/03 virus did not result in increased pathogenicity in mice (Munster *et al.*, 2007).

Recent studies have sought to identify molecular correlates of virulence within the NS1 gene. A five amino acid deletion in the NS1 gene (positions 80–84) found in recently isolated H5N1 viruses has been associated with higher viral titers, delayed viral clearance, and increased lethality following infection in mice compared with H5N1 viruses that do not possess this deletion or mutant viruses that possessed an artificial insertion of these amino acids (Li *et al.*, 2004; Long *et al.*, 2008). Additionally, a single amino acid substitution in the NS1 protein of the H5N1 virus A/Duck/Guangxi/12/03 was found to correlate with high and low pathogenicity phenotypes in mice (Jiao *et al.*, 2008). This substitution, P42S, resulted in systemic spread and increased virulence of the H5N1 virus following infection in mice as compared with wild-type virus which contained a proline at this position. Sequence from the NS1 protein extreme C terminus of HPAI H5N1 viruses was recently shown to enhance the pathogenicity of the human virus A/WSN/33 in mice, identifying this region as a molecular determinant of pathogenicity (Jackson *et al.*, 2008). Further studies are necessary to elucidate the complex role of NS1 during influenza virus infection, as well as identify possible molecular targets to circumvent this inhibition of host immune responses for therapeutic development.

VII. CONCLUSIONS

The use of mammalian models to study influenza virus infection has revealed extensive information regarding virus–host interactions that is essential for the full understanding of avian influenza virus pathogenesis. As we have detailed in this review, both viral and host factors contribute to the overall pathogenicity of avian influenza viruses. As avian influenza viruses continue to pose a major public health threat, this work advances the overall understanding of H5 and H7 influenza viruses associated with disease in humans and offers many further avenues of investigation to study the pathogenesis of avian influenza viruses of multiple subtypes in the context of human infection. Further study of viruses within both subtypes is warranted to better understand the properties that confer a high pathogenicity phenotype in mammals and to best prepare for future pandemics. Such research is being done with the hope that the knowledge gained will allow the world to better prepare for and respond to future influenza pandemics.

Developing vaccination strategies for avian influenza viruses has become increasingly important as these viruses continue to cause outbreaks in poultry. Animal models of H5 and H7 influenza infection have served as a useful tool in assessing not only the relative virulence of a virus but also in assessing the suitability of a particular virus for vaccine development (Joseph *et al.*, 2007). With the use of animal models, reassortant vaccine candidates targeting H5 and H7 viruses have been generated and await human testing (de Wit *et al.*, 2005; Joseph *et al.*, 2008; Pappas *et al.*, 2007; Subbarao and Luke, 2007). As the subtype of future pandemics cannot be known in advance, mouse and ferret models have further been established for other avian virus subtypes, including H9 and H6 viruses (Gillim-Ross *et al.*, 2008; Guo *et al.*, 2000; Wan *et al.*, 2008). These models have additionally served a vital purpose for the evaluation of many novel vaccination strategies toward human and avian influenza, including the use of adjuvants, virus-like particles, adenoviral vectors, transdermal delivery systems, and others (Bright *et al.*, 2008; Garg *et al.*, 2007; Hoelscher *et al.*, 2006; Lu *et al.*, 2006).

While vaccination offers the best protection against influenza, the 6–8-month timeframe necessary to manufacture an antigenically well-matched vaccine against a novel virus suggests that antivirals, and not vaccines, will be the most readily available first line of defense against a pandemic (Stephenson *et al.*, 2004). There are currently two classes of influenza antiviral drugs available for human use: the M2 ion channel blockers (amantadine and rimantadine) and the neuraminidase inhibitors (oseltamivir and zanamivir) (Couch, 2000). Using mouse and ferret models, both classes of antiviral drugs have been shown to be effective against avian influenza viruses (Govorkova *et al.*, 2001, 2007; Gubareva *et al.*, 1998). Studies evaluating the emergence of drug-resistant mutants following antiviral treatment have also been possible with the use of these *in vivo* models (Gubareva *et al.*, 1998; Yen *et al.*, 2005b). Accordingly, mammalian models have also been used to evaluate the efficacy of new treatments against avian influenza viruses (Belser *et al.*, 2007b; Tompkins *et al.*, 2004).

HPAI A viruses continue to cause extensive outbreaks of disease in domestic and wild birds and an ever increasing number of human infections and fatalities. Within the past decade, highly pathogenic H5N1 viruses have become endemic in some Southeast Asian countries and have spread to the Middle East, Europe, and Africa. During this time, viruses of high and low pathogenicity within the H7 subtype have additionally resulted in over 100 cases of human infection, primarily in Europe but also in North America. HPAI viruses pose a considerable threat to public health as they constitute a novel HA subtype emerging in a serologically naïve human population. However, they currently lack efficient person-to-person transmissibility, a critical parameter for pandemic virus strains. Pandemics of the twentieth century varied substantially in their severity, and the public health response, including continuous investigation of strategies and recommendations for early vaccine coverage, will depend on the severity of the next pandemic. Therefore, understanding the mechanisms of virulence of avian influenza viruses is crucial not only to develop improved treatment options for clinical care but also as a means to estimate the likely severity of disease for a given pandemic strain. The research detailed within this review demonstrates the great strides made toward understanding those molecular determinants of avian influenza viruses that confer a highly pathogenic phenotype in mammals. Additional studies that link this work with those identified virologic properties associated with previous pandemic strains will further our ability to predict those viruses with pandemic potential and prevent infection by means of vaccines and antivirals. As these viruses will continue to evolve in the years to come, it is essential to continue these avenues of investigation so that we are best prepared to combat a future pandemic.

ACKNOWLEDGMENTS

We thank our colleagues within international public health agencies and members of the WHO Influenza Network for making available many of the viruses discussed in this review, including Nguyen Tran Hien and Le Quynh Mai (Vietnam Ministry of Health), Pranee Thawatsupha, Malinee Chittaganpitch, and Sunthareeya Waicharoen (Thai National Institute of Health), Wilina Lim (Hong Kong Department of Health), David Swayne (Southeast Poultry Research Laboratory, United States Department of Agriculture, Agricultural Research Service), Ron Fouchier (Erasmus Medical Center), and Yan Li (Canadian Center for Human and Animal Health). The findings and conclusions in this report are those of the authors and do not necessarily represent the views of the funding agency.

REFERENCES

Abdel-Ghafar, A. N., Chotpitayasunondh, T., Gao, Z., Hayden, F. G., Nguyen, D. H., de Jong, M. D., Naghdaliyev, A., Peiris, J. S., Shindo, N., Soeroso, S., and Uyeki, T. M. (2008). Update on avian influenza A (H5N1) virus infection in humans. *N. Engl. J. Med.* **358**(3):261–273.

Akkina, R. K., Chambers, T. M., Londo, D. R., and Nayak, D. P. (1987). Intracellular localization of the viral polymerase proteins in cells infected with influenza virus and cells expressing PB1 protein from cloned cDNA. *J. Virol.* **61**(7):2217–2224.

Alford, R. H., Kasel, J. A., Gerone, P. J., and Knight, V. (1966). Human influenza resulting from aerosol inhalation. *Proc. Soc. Exp. Biol. Med.* **122**(3):800–804.

Almond, J. W. (1977). A single gene determines the host range of influenza virus. *Nature* **270** (5638):617–618.

Apisarnthanarak, A., Kitphati, R., Thongphubeth, K., Patoomanunt, P., Anthanont, P., Auwanit, W., Thawatsupha, P., Chittaganpitch, M., Saeng-Aroon, S., Waicharoen, S., Apisarnthanarak, P., Storch, G. A., *et al.* (2004). Atypical avian influenza (H5N1). *Emerg. Infect. Dis.* **10**(7):1321–1324.

Baigent, S. J., Bethell, R. C., and McCauley, J. W. (1999). Genetic analysis reveals that both haemagglutinin and neuraminidase determine the sensitivity of naturally occurring avian influenza viruses to zanamivir *in vitro. Virology* **263**(2):323–338.

Baigent, S. J., and McCauley, J. W. (2001). Glycosylation of haemagglutinin and stalk-length of neuraminidase combine to regulate the growth of avian influenza viruses in tissue culture. *Virus Res.* **79**(1–2):177–185.

Banks, J., Speidel, E. S., Moore, E., Plowright, L., Piccirillo, A., Capua, I., Cordioli, P., Fioretti, A., and Alexander, D. J. (2001). Changes in the haemagglutinin and the neuraminidase genes prior to the emergence of highly pathogenic H7N1 avian influenza viruses in Italy. *Arch. Virol.* **146**(5):963–973.

Basarab, O., and Smith, H. (1969). Quantitative studies on the tissue localization of influenza virus in ferrets after intranasal and intravenous or intracardial inoculation. *Br. J. Exp. Pathol.* **50**(6):612–618.

Bean, B., Moore, B. M., Sterner, B., Peterson, L. R., Gerding, D. N., and Balfour, H. H., Jr. (1982). Survival of influenza viruses on environmental surfaces. *J. Infect. Dis.* **146**(1):47–51.

Beare, A. S., and Webster, R. G. (1991). Replication of avian influenza viruses in humans. *Arch. Virol.* **119**(1–2):37–42.

Belser, J. A., Blixt, O., Chen, L. M., Pappas, C., Maines, T. R., Van Hoeven, N., Donis, R., Busch, J., McBride, R., Paulson, J. C., Katz, J. M., and Tumpey, T. M. (2008). Contemporary North American Influenza H7 viruses possess human receptor specificity: Implications for virus transmissibility. *Proc. Natl. Acad. Sci. USA* **105**(21):7558–7563.

Belser, J. A., Lu, X., Maines, T. R., Smith, C., Li, Y., Donis, R. O., Katz, J. M., and Tumpey, T. M. (2007a). Pathogenesis of avian influenza (H7) virus infection in mice and ferrets: Enhanced virulence of Eurasian H7N7 viruses isolated from humans. *J. Virol.* **81**(20):11139–11147.

Belser, J. A., Lu, X., Szretter, K. J., Jin, X., Aschenbrenner, L. M., Lee, A., Hawley, S., Kimdo, H., Malakhov, M. P., Yu, M., Fang, F., and Katz, J. M. (2007b). DAS181, a novel sialidase fusion protein, protects mice from lethal avian influenza H5N1 virus infection. *J. Infect. Dis.* **196**(10):1493–1499.

Belser, J. A., Wadford, D. A., Xu, J., Katz, J. M., and Tumpey, T. M. (2009). Ocular infection of mice with influenza A (H7) viruses: A site of primary replication and spread to the respiratory tract. *J. Virol.* **83**(14):7075–7084.

Bosch, F. X., Garten, W., Klenk, H. D., and Rott, R. (1981). Proteolytic cleavage of influenza virus hemagglutinins: Primary structure of the connecting peptide between HA1 and HA2 determines proteolytic cleavability and pathogenicity of Avian influenza viruses. *Virology* **113**(2):725–735.

Bottcher, E., Matrosovich, T., Beyerle, M., Klenk, H. D., Garten, W., and Matrosovich, M. (2006). Proteolytic activation of influenza viruses by serine proteases TMPRSS2 and HAT from human airway epithelium. *J. Virol.* **80**(19):9896–9898.

Bridges, C. B., Kuehnert, M. J., and Hall, C. B. (2003). Transmission of influenza: Implications for control in health care settings. *Clin. Infect. Dis.* **37**(8):1094–1101.

Bright, R. A., Carter, D. M., Crevar, C. J., Toapanta, F. R., Steckbeck, J. D., Cole, K. S., Kumar, N. M., Pushko, P., Smith, G., Tumpey, T. M., and Ross, T. M. (2008). Cross-Clade Protective Immune Responses to Influenza Viruses with H5N1 HA and NA Elicited by an Influenza Virus-Like Particle. *PLoS ONE* **3**(1):e1501.

Bright, R. A., Cho, D. S., Rowe, T., and Katz, J. M. (2003). Mechanisms of pathogenicity of influenza A (H5N1) viruses in mice. *Avian Dis.* **47**(Suppl. 3):1131–1134.

Brown, E. G., and Bailly, J. E. (1999). Genetic analysis of mouse-adapted influenza A virus identifies roles for the NA, PB1, and PB2 genes in virulence. *Virus Res.* **61**(1):63–76.

Brown, E. G., Liu, H., Kit, L. C., Baird, S., and Nesrallah, M. (2001). Pattern of mutation in the genome of influenza A virus on adaptation to increased virulence in the mouse lung: Identification of functional themes. *Proc. Natl. Acad. Sci. USA* **98**(12):6883–6888.

Butt, K. M., Smith, G. J., Chen, H., Zhang, L. J., Leung, Y. H., Xu, K. M., Lim, W., Webster, R. G., Yuen, K. Y., Peiris, J. S., and Guan, Y. (2005). Human Infection with an Avian H9N2 Influenza A Virus in Hong Kong in 2003. *J. Clin. Microbiol.* **43**(11):5760–5767.

Capua, I., and Alexander, D. J. (2004). Avian influenza: Recent developments. *Avian Pathol.* **33**(4):393–404.

Castrucci, M. R., and Kawaoka, Y. (1993). Biologic importance of neuraminidase stalk length in influenza A virus. *J. Virol.* **67**(2):759–764.

Cavanagh, D., Mitkis, F., Sweet, C., Collie, M. H., and Smith, H. (1979). The localization of influenza virus in the respiratory tract of ferrets: Susceptible nasal mucosa cells produce and release more virus than susceptible lung cells. *J. Gen. Virol.* **44**(2):505–514.

CDC (2004a). Cases of influenza A (H5N1) - Thailand 2004 . *Morb. Mortal. Wkly Rep.* **53**:100–103.

CDC (2004b). Update: Influenza activity—United States and worldwide, 2003–04 season, and composition of the 2004–05 influenza vaccine. *MMWR Morb. Mortal. Wkly Rep.* **53**(25):547–552.

CDC (2004c). Update: Influenza activity—United States, 2003–04 season. *MMWR Morb. Mortal. Wkly Rep.* **53**(13):284–287.

CDC (2006). *Interim Guidance for Protection of Persons Involved in U.S. Avian Influenza Outbreak Disease Control and Eradication Activities.* Health and Human Services.

Chandrasekaran, A., Srinivasan, A., Raman, R., Viswanathan, K., Raguram, S., Tumpey, T. M., Sasisekharan, V., and Sasisekharan, R. (2008). Glycan topology determines human adaptation of avian H5N1 virus hemagglutinin. *Nat. Biotechnol.* **26**(1):107–113.

Chen, H., Bright, R. A., Subbarao, K., Smith, C., Cox, N. J., Katz, J. M., and Matsuoka, Y. (2007). Polygenic virulence factors involved in pathogenesis of 1997 Hong Kong H5N1 influenza viruses in mice. *Virus Res.* **128**(1–2):159–163.

Chen, L. M., Davis, C. T., Zhou, H., Cox, N. J., and Donis, R. O. (2008). Genetic compatibility and virulence of reassortants derived from contemporary avian H5N1 and human H3N2 influenza A viruses. *PLoS Pathog.* **4**(5):e1000072.

Chen, W., Calvo, P. A., Malide, D., Gibbs, J., Schubert, U., Bacik, I., Basta, S., O'Neill, R., Schickli, J., Palese, P., Henklein, P., Bennink, J. R., *et al.* (2001). A novel influenza A virus mitochondrial protein that induces cell death. *Nat. Med.* **7**(12):1306–1312.

Chen, Z., and Krug, R. M. (2000). Selective nuclear export of viral mRNAs in influenza-virus-infected cells. *Trends Microbiol.* **8**(8):376–383.

Chen, Z., Li, Y., and Krug, R. M. (1999). Influenza A virus NS1 protein targets poly(A)-binding protein II of the cellular 3′-end processing machinery. *EMBO J.* **18**(8):2273–2283.

Cheung, C. Y., Poon, L. L., Lau, A. S., Luk, W., Lau, Y. L., Shortridge, K. F., Gordon, S., Guan, Y., and Peiris, J. S. (2002). Induction of proinflammatory cytokines in human macrophages by influenza A (H5N1) viruses: A mechanism for the unusual severity of human disease? *Lancet* **360**(9348):1831–1837.

Claas, E. C., Osterhaus, A. D., van Beek, R., De Jong, J. C., Rimmelzwaan, G. F., Senne, D. A., Krauss, S., Shortridge, K. F., and Webster, R. G. (1998). Human influenza A H5N1 virus related to a highly pathogenic avian influenza virus. *Lancet* **351**(9101):472–477.

Coleman, J. R. (2007). The PB1–F2 protein of Influenza A virus: Increasing pathogenicity by disrupting alveolar macrophages. *J. Virol.* **4**:9.

Conenello, G. M., Zamarin, D., Perrone, L. A., Tumpey, T., and Palese, P. (2007). A single mutation in the PB1–F2 of H5N1 (HK/97) and 1918 influenza A viruses contributes to increased virulence. *PLoS Pathog.* **3**(10):1414–1421.

Conn, C. A., McClellan, J. L., Maassab, H. F., Smitka, C. W., Majde, J. A., and Kluger, M. J. (1995). Cytokines and the acute phase response to influenza virus in mice. *Am. J. Physiol.* **268**(1 Pt 2):R78–R84.

Couch, R. B. (2000). Prevention and treatment of influenza. *N. Engl. J. Med.* **343**(24):1778–1787.

Crescenzo-Chaigne, B., van der Werf, S., and Naffakh, N. (2002). Differential effect of nucleotide substitutions in the 3' arm of the influenza A virus vRNA promoter on transcription/replication by avian and human polymerase complexes is related to the nature of PB2 amino acid 627. *Virology* **303**(2):240–252.

de Jong, J. C., Claas, E. C., Osterhaus, A. D., Webster, R. G., and Lim, W. L. (1997). A pandemic warning? *Nature* **389**(6651):554.

de Jong, M. D., Bach, V. C., Phan, T. Q., Vo, M. H., Tran, T. T., Nguyen, B. H., Beld, M., Le, T. P., Truong, H. K., Nguyen, V. V., Tran, T. H., Do, Q. H., *et al.* (2005). Fatal avian influenza A (H5N1) in a child presenting with diarrhea followed by coma. *N. Engl. J. Med.* **352**(7):686–691.

de Jong, M. D., Simmons, C. P., Thanh, T. T., Hien, V. M., Smith, G. J., Chau, T. N., Hoang, D. M., Chau, N. V., Khanh, T. H., Dong, V. C., Qui, P. T., Cam, B. V., *et al.* (2006). Fatal outcome of human influenza A (H5N1) is associated with high viral load and hypercytokinemia. *Nat. Med.* **12**(10):1203–1207.

de Wit, E., Munster, V. J., Spronken, M. I., Bestebroer, T. M., Baas, C., Beyer, W. E., Rimmelzwaan, G. F., Osterhaus, A. D., and Fouchier, R. A. (2005). Protection of mice against lethal infection with highly pathogenic H7N7 influenza A virus by using a recombinant low-pathogenicity vaccine strain. *J. Virol.* **79**(19):12401–12407.

Deshpande, K. L., Fried, V. A., Ando, M., and Webster, R. G. (1987). Glycosylation affects cleavage of an H5N2 influenza virus hemagglutinin and regulates virulence. *Proc. Natl. Acad. Sci. USA* **84**(1):36–40.

Doherty, P. C., Allan, W., Eichelberger, M., and Carding, S. R. (1992). Roles of alpha beta and gamma delta T cell subsets in viral immunity. *Annu. Rev. Immunol.* **10**:123–151.

Droebner, K., Reiling, S. J., and Planz, O. (2008). Role of hypercytokinemia in NF-kappaB p50-deficient mice after H5N1 influenza A virus infection. *J. Virol.* **82**(22):11461–11466.

Dudley, J. P. (2008). Public Health and Epidemiological Considerations for Avian Influenza Risk Mapping and Risk Assessment. *Ecol. Soc.* **13**(2):21.

Dupuis, S., Jouanguy, E., Al-Hajjar, S., Fieschi, C., Al-Mohsen, I. Z., Al-Jumaah, S., Yang, K., Chapgier, A., Eidenschenk, C., Eid, P., Al Ghonaium, A., Tufenkeji, H., *et al.* (2003). Impaired response to interferon-alpha/beta and lethal viral disease in human STAT1 deficiency. *Nat. Genet.* **33**(3):388–391.

Easterday, B. C. (1975). Animal Influenza. *In* "The influenza viruses and influenza", (E. D. Kilbourne, Ed.), pp. 449–481. Academic Press, Orlando.

Eurosurveillance Editorial Team (2007). Avian influenza A/(H7N2) outbreak in the United Kingdom . *Euro Surveill.* **12**(5):E070531.2.

Fodor, E., Devenish, L., Engelhardt, O. G., Palese, P., Brownlee, G. G., and Garcia-Sastre, A. (1999). Rescue of influenza A virus from recombinant DNA. *J. Virol.* **73**(11):9679–9682.

Fortes, P., Beloso, A., and Ortin, J. (1994). Influenza virus NS1 protein inhibits pre-mRNA splicing and blocks mRNA nucleocytoplasmic transport. *EMBO J.* **13**(3):704–712.

Fouchier, R. A., Munster, V., Wallensten, A., Bestebroer, T. M., Herfst, S., Smith, D., Rimmelzwaan, G. F., Olsen, B., and Osterhaus, A. D. (2005). Characterization of a novel influenza A virus hemagglutinin subtype (H16) obtained from black-headed gulls. *J. Virol.* **79**(5):2814–2822.

Fouchier, R. A., Schneeberger, P. M., Rozendaal, F. W., Broekman, J. M., Kemink, S. A., Munster, V., Kuiken, T., Rimmelzwaan, G. F., Schutten, M., Van Doornum, G. J., Koch, G., Bosman, A., *et al.* (2004). Avian influenza A virus (H7N7) associated with human conjunctivitis and a fatal case of acute respiratory distress syndrome. *Proc. Natl. Acad. Sci. USA* **101**(5):1356–1361.

Gabriel, G., Dauber, B., Wolff, T., Planz, O., Klenk, H. D., and Stech, J. (2005). The viral polymerase mediates adaptation of an avian influenza virus to a mammalian host. *Proc. Natl. Acad. Sci. USA* **102**(51):18590–18595.

Gambaryan, A. S., Tuzikov, A. B., Piskarev, V. E., Yamnikova, S. S., Lvov, D. K., Robertson, J. S., Bovin, N. V., and Matrosovich, M. N. (1997). Specification of receptor-binding phenotypes of influenza virus isolates from different hosts using synthetic sialylglycopolymers: Non-egg-adapted human H1 and H3 influenza A and influenza B viruses share a common high binding affinity for 6'-sialyl(N-acetyllactosamine). *Virology* **232**(2):345–350.

Gao, P., Watanabe, S., Ito, T., Goto, H., Wells, K., McGregor, M., Cooley, A. J., and Kawaoka, Y. (1999). Biological heterogeneity, including systemic replication in mice, of H5N1 influenza A virus isolates from humans in Hong Kong. *J. Virol.* **73**(4):3184–3189.

Garcia-Sastre, A. (2001). Inhibition of interferon-mediated antiviral responses by influenza A viruses and other negative-strand RNA viruses. *Virology* **279**(2):375–384.

Garg, S., Hoelscher, M., Belser, J. A., Wang, C., Jayashankar, L., Guo, Z., Durland, R. H., Katz, J. M., and Sambhara, S. (2007). Needle-free skin patch delivery of a vaccine for a potentially pandemic influenza virus provides protection against lethal challenge in mice. *Clin. Vaccine Immunol.* **14**(7):926–928.

Geiss, G. K., Salvatore, M., Tumpey, T. M., Carter, V. S., Wang, X., Basler, C. F., Taubenberger, J. K., Bumgarner, R. E., Palese, P., Katze, M. G., and Garcia-Sastre, A. (2002). Cellular transcriptional profiling in influenza A virus-infected lung epithelial cells: The role of the nonstructural NS1 protein in the evasion of the host innate defense and its potential contribution to pandemic influenza. *Proc. Natl. Acad. Sci. USA* **99**(16):10736–10741.

Gibbs, M. J., and Gibbs, A. J. (2006). Molecular virology: Was the 1918 pandemic caused by a bird flu? *Nature* **440**(7088):E8 (discussion E9–10).

Giese, M., Harder, T. C., Teifke, J. P., Klopfleisch, R., Breithaupt, A., Mettenleiter, T. C., and Vahlenkamp, T. W. (2008). Experimental infection and natural contact exposure of dogs with avian influenza virus (H5N1). *Emerg. Infect. Dis.* **14**(2):308–310.

Gillim-Ross, L., Santos, C., Chen, Z., Aspelund, A., Yang, C. F., Ye, D., Jin, H., Kemble, G., and Subbarao, K. (2008). Avian influenza h6 viruses productively infect and cause illness in mice and ferrets. *J. Virol.* **82**(21):10854–10863.

Goto, H., and Kawaoka, Y. (1998). A novel mechanism for the acquisition of virulence by a human influenza A virus. *Proc. Natl. Acad. Sci. USA* **95**(17):10224–10228.

Goto, H., Wells, K., Takada, A., and Kawaoka, Y. (2001). Plasminogen-binding activity of neuraminidase determines the pathogenicity of influenza A virus. *J. Virol.* **75**(19):9297–9301.

Govorkova, E. A., Ilyushina, N. A., Boltz, D. A., Douglas, A., Yilmaz, N., and Webster, R. G. (2007). Efficacy of oseltamivir therapy in ferrets inoculated with different clades of H5N1 influenza virus. *Antimicrob. Agents Chemother.* **51**(4):1414–1424.

Govorkova, E. A., Leneva, I. A., Goloubeva, O. G., Bush, K., and Webster, R. G. (2001). Comparison of efficacies of RWJ-270201, zanamivir, and oseltamivir against H5N1, H9N2, and other avian influenza viruses. *Antimicrob. Agents Chemother.* **45**(10):2723–2732.

Govorkova, E. A., Rehg, J. E., Krauss, S., Yen, H. L., Guan, Y., Peiris, M., Nguyen, T. D., Hanh, T. H., Puthavathana, P., Long, H. T., Buranathai, C., Lim, W., *et al.* (2005). Lethality to ferrets of H5N1 influenza viruses isolated from humans and poultry in 2004. *J. Virol.* **79**(4):2191–2198.

Guan, Y., Poon, L. L., Cheung, C. Y., Ellis, T. M., Lim, W., Lipatov, A. S., Chan, K. H., Sturm-Ramirez, K. M., Cheung, C. L., Leung, Y. H., Yuen, K. Y., Webster, R. G., *et al.* (2004). H5N1 influenza: A protean pandemic threat. *Proc. Natl. Acad. Sci. USA* **101**(21):8156–8161.

Gubareva, L. V., McCullers, J. A., Bethell, R. C., and Webster, R. G. (1998). Characterization of influenza A/HongKong/156/97 (H5N1) virus in a mouse model and protective effect of zanamivir on H5N1 infection in mice. *J. Infect. Dis.* **178**(6):1592–1596.

Guo, Y. J., Krauss, S., Senne, D. A., Mo, I. P., Lo, K. S., Xiong, X. P., Norwood, M., Shortridge, K. F., Webster, R. G., and Guan, Y. (2000). Characterization of the pathogenicity of members of the newly established H9N2 influenza virus lineages in Asia. *Virology* **267**(2):279–288.

Guo, Z., Chen, L. M., Zeng, H., Gomez, J. A., Plowden, J., Fujita, T., Katz, J. M., Donis, R. O., and Sambhara, S. (2007). NS1 protein of influenza A virus inhibits the function of intracytoplasmic pathogen sensor, RIG-I. *Am. J. Respir. Cell Mol. Biol.* **36**(3):263–269.

Haff, R. F., Schriver, P. W., Engle, C. G., and Stewart, R. C. (1966). Pathogenesis of influenza in ferrets. I. Tissue and blood manifestations of disease. *J. Immunol.* **96**(4):659–667.

Haller, O., Kochs, G., and Weber, F. (2006). The interferon response circuit: Induction and suppression by pathogenic viruses. *Virology* **344**(1):119–130.

Hatta, M., Gao, P., Halfmann, P., and Kawaoka, Y. (2001). Molecular basis for high virulence of Hong Kong H5N1 influenza A viruses. *Science* **293**(5536):1840–1842.

Hatta, M., Hatta, Y., Kim, J. H., Watanabe, S., Shinya, K., Nguyen, T., Lien, P. S., Le, Q. M., and Kawaoka, Y. (2007). Growth of H5N1 influenza A viruses in the upper respiratory tracts of mice. *PLoS Pathog.* **3**(10):1374–1379.

Hennet, T., Ziltener, H. J., Frei, K., and Peterhans, E. (1992). A kinetic study of immune mediators in the lungs of mice infected with influenza A virus. *J. Immunol.* **149**(3):932–939.

Herlocher, M. L., Elias, S., Truscon, R., Harrison, S., Mindell, D., Simon, C., and Monto, A. S. (2001). Ferrets as a transmission model for influenza: Sequence changes in HA1 of type A (H3N2) virus. *J. Infect. Dis.* **184**(5):542–546.

Hinshaw, V. S., Webster, R. G., Easterday, B. C., and Bean, W. J., Jr. (1981). Replication of avian influenza A viruses in mammals. *Infect. Immun.* **34**(2):354–361.

Hirst, M., Astell, C. R., Griffith, M., Coughlin, S. M., Moksa, M., Zeng, T., Smailus, D. E., Holt, R. A., Jones, S., Marra, M. A., Petric, M., Krajden, M., *et al.* (2004). Novel avian influenza H7N3 strain outbreak, British Columbia. *Emerg. Infect. Dis.* **10**(12):2192–2195.

Hoelscher, M. A., Garg, S., Bangari, D. S., Belser, J. A., Lu, X., Stephenson, I., Bright, R. A., Katz, J. M., Mittal, S. K., and Sambhara, S. (2006). Development of adenoviral-vector-based pandemic influenza vaccine against antigenically distinct human H5N1 strains in mice. *Lancet* **367**(9509):475–481.

Horimoto, T., Rivera, E., Pearson, J., Senne, D., Krauss, S., Kawaoka, Y., and Webster, R. G. (1995). Origin and molecular changes associated with emergence of a highly pathogenic H5N2 influenza virus in Mexico. *Virology* **213**(1):223–230.

Hulse, D. J., Webster, R. G., Russell, R. J., and Perez, D. R. (2004). Molecular determinants within the surface proteins involved in the pathogenicity of H5N1 influenza viruses in chickens. *J. Virol.* **78**(18):9954–9964.

Ibricevic, A., Pekosz, A., Walter, M. J., Newby, C., Battaile, J. T., Brown, E. G., Holtzman, M. J., and Brody, S. L. (2006). Influenza virus receptor specificity and cell tropism in mouse and human airway epithelial cells. *J. Virol.* **80**(15):7469–7480.

Ito, T., and Kawaoka, Y. (2000). Host-range barrier of influenza A viruses. *Vet. Microbiol.* **74**(1–2):71–75.

Jackson, D., Hossain, M. J., Hickman, D., Perez, D. R., and Lamb, R. A. (2008). A new influenza virus virulence determinant: The NS1 protein four C-terminal residues modulate pathogenicity. *Proc. Natl. Acad. Sci. USA* **105**(11):4381–4386.

Jiao, P., Tian, G., Li, Y., Deng, G., Jiang, Y., Liu, C., Liu, W., Bu, Z., Kawaoka, Y., and Chen, H. (2008). A single-amino-acid substitution in the NS1 protein changes the pathogenicity of H5N1 avian influenza viruses in mice. *J. Virol.* **82**(3):1146–1154.

Johnson, N. P., and Mueller, J. (2002). Updating the accounts: Global mortality of the 1918–1920 "Spanish" influenza pandemic. *Bull. Hist. Med.* **76**(1):105–115.

Joseph, T., McAuliffe, J., Lu, B., Jin, H., Kemble, G., and Subbarao, K. (2007). Evaluation of replication and pathogenicity of avian influenza a H7 subtype viruses in a mouse model. *J. Virol.* **81**(19):10558–10566.

Joseph, T., McAuliffe, J., Lu, B., Vogel, L., Swayne, D., Jin, H., Kemble, G., and Subbarao, K. (2008). A live attenuated cold-adapted influenza A H7N3 virus vaccine provides protection against homologous and heterologous H7 viruses in mice and ferrets. *Virology* **378**(1):123–132.

Kandun, I. N., Wibisono, H., Sedyaningsih, E. R., Yusharmen, B., Hadisoedarsuno, W., Purba, W., Santoso, H., Septiawati, C., Tresnaningsih, E., Heriyanto, B., Yuwono, D., Harun, S., *et al.* (2006). Three Indonesian clusters of H5N1 virus infection in 2005. *N. Engl. J. Med.* **355**(21):2186–2194.

Katz, J. M., Lu, X., Frace, A. M., Morken, T., Zaki, S. R., and Tumpey, T. M. (2000a). Pathogenesis of and immunity to avian influenza A H5 viruses. *Biomed. Pharmacother.* **54**(4):178–187.

Katz, J. M., Lu, X., Tumpey, T. M., Smith, C. B., Shaw, M. W., and Subbarao, K. (2000b). Molecular correlates of influenza A H5N1 virus pathogenesis in mice. *J. Virol.* **74**(22):10807–10810.

Kawaoka, Y. (1991). Equine H7N7 influenza A viruses are highly pathogenic in mice without adaptation: Potential use as an animal model. *J. Virol.* **65**(7):3891–3894.

Kawaoka, Y., Bordwell, E., and Webster, R. G. (1987). Intestinal replication of influenza A viruses in two mammalian species. Brief report. *Arch Virol.* **93**(3–4):303–308.

Kawaoka, Y., Krauss, S., and Webster, R. G. (1989). Avian-to-human transmission of the PB1 gene of influenza A viruses in the 1957 and 1968 pandemics. *J. Virol.* **63**(11):4603–4608.

Koopmans, M., Wilbrink, B., Conyn, M., Natrop, G., van der Nat, H., Vennema, H., Meijer, A., van Steenbergen, J., Fouchier, R., Osterhaus, A., and Bosman, A. (2004). Transmission of H7N7 avian influenza A virus to human beings during a large outbreak in commercial poultry farms in The Netherlands. *Lancet* **363**(9409):587–593.

Kurokawa, M., Imakita, M., Kumeda, C. A., and Shiraki, K. (1996). Cascade of fever production in mice infected with influenza virus. *J. Med. Virol.* **50**(2):152–158.

Kurtz, J., Manvell, R. J., and Banks, J. (1996). Avian influenza virus isolated from a woman with conjunctivitis. *Lancet* **348**(9031):901–902.

Lamb, R. A., and Choppin, P. W. (1976). Synthesis of influenza virus proteins in infected cells: Translation of viral polypeptides, including three P polypeptides, from RNA produced by primary transcription. *Virology* **74**(2):504–519.

Lee, C. W., Lee, Y. J., Senne, D. A., and Suarez, D. L. (2006). Pathogenic potential of North American H7N2 avian influenza virus: A mutagenesis study using reverse genetics. *Virology* **353**(2):388–395.

Leigh, M. W., Connor, R. J., Kelm, S., Baum, L. G., and Paulson, J. C. (1995). Receptor specificity of influenza virus influences severity of illness in ferrets. *Vaccine* **13**(15):1468–1473.

Li, K. S., Guan, Y., Wang, J., Smith, G. J., Xu, K. M., Duan, L., Rahardjo, A. P., Puthavathana, P., Buranathai, C., Nguyen, T. D., Estoepangestie, A. T., Chaisingh, A., *et al.* (2004). Genesis of a highly pathogenic and potentially pandemic H5N1 influenza virus in eastern Asia. *Nature* **430**(6996):209–213.

Li, S., Schulman, J., Itamura, S., and Palese, P. (1993). Glycosylation of neuraminidase determines the neurovirulence of influenza A/WSN/33 virus. *J. Virol.* **67**(11):6667–6673.

Li, Z., Chen, H., Jiao, P., Deng, G., Tian, G., Li, Y., Hoffmann, E., Webster, R. G., Matsuoka, Y., and Yu, K. (2005). Molecular basis of replication of duck H5N1 influenza viruses in a mammalian mouse model. *J. Virol.* **79**(18):12058–12064.

Lidwell, O. M. (1974). Aerial dispersal of micro-organisms from the human respiratory tract. *Soc. Appl. Bacteriol. Symp. Ser.* **3:**135–154.

Lipatov, A. S., Kwon, Y. K., Pantin-Jackwood, M. J., and Swayne, D. E. (2009). Pathogenesis of H5N1 influenza virus infections in mice and ferret models differs according to respiratory tract or digestive system exposure. *J. Infect. Dis.* **199**(5):717–725.

Long, J. X., Peng, D. X., Liu, Y. L., Wu, Y. T., and Liu, X. F. (2008). Virulence of H5N1 avian influenza virus enhanced by a 15-nucleotide deletion in the viral nonstructural gene. *Virus Genes* **36**(3):471–478.

Lowen, A. C., Mubareka, S., Steel, J., and Palese, P. (2007). Influenza virus transmission is dependent on relative humidity and temperature. *PLoS Pathog.* **3**(10):1470–1476.

Lowen, A. C., Mubareka, S., Tumpey, T. M., Garcia-Sastre, A., and Palese, P. (2006). The guinea pig as a transmission model for human influenza viruses. *Proc. Natl. Acad. Sci. USA* **103**(26):9988–9992.

Lowen, A. C., Steel, J., Mubareka, S., and Palese, P. (2008). High temperature (30 degrees C) blocks aerosol but not contact transmission of influenza virus. *J. Virol.* **82**(11):5650–5652.

Lu, X., Cho, D., Hall, H., Rowe, T., Sung, H., Kim, W., Kang, C., Mo, I., Cox, N., Klimov, A., and Katz, J. (2003). Pathogenicity and antigenicity of a new influenza A (H5N1) virus isolated from duck meat. *J. Med. Virol.* **69**(4):553–559.

Lu, X., Edwards, L. E., Desheva, J. A., Nguyen, D. C., Rekstin, A., Stephenson, I., Szretter, K., Cox, N. J., Rudenko, L. G., Klimov, A., and Katz, J. M. (2006). Cross-protective immunity in mice induced by live-attenuated or inactivated vaccines against highly pathogenic influenza A (H5N1) viruses. *Vaccine* **24**(44–46):6588–6593.

Lu, X., Tumpey, T. M., Morken, T., Zaki, S. R., Cox, N. J., and Katz, J. M. (1999). A mouse model for the evaluation of pathogenesis and immunity to influenza A (H5N1) viruses isolated from humans. *J. Virol.* **73**(7):5903–5911.

Ludwig, S., Pleschka, S., and Wolff, T. (1999). A fatal relationship—influenza virus interactions with the host cell. *Viral Immunol.* **12**(3):175–196.

Ma, W., Vincent, A. L., Gramer, M. R., Brockwell, C. B., Lager, K. M., Janke, B. H., Gauger, P. C., Patnayak, D. P., Webby, R. J., and Richt, J. A. (2007). Identification of H2N3 influenza A viruses from swine in the United States. *Proc. Natl. Acad. Sci. USA* **104** (52):20949–20954.

Maher, J. A., and DeStefano, J. (2004). The ferret: An animal model to study influenza virus. *Lab. Anim. (NY)* **33**(9):50–53.

Maines, T. R., Chen, L. M., Matsuoka, Y., Chen, H., Rowe, T., Ortin, J., Falcon, A., Nguyen, T. H., Mai le, Q., Sedyaningsih, E. R., Harun, S., Tumpey, T. M., *et al.* (2006). Lack of transmission of H5N1 avian–human reassortant influenza viruses in a ferret model. *Proc. Natl. Acad. Sci. USA* **103**(32):12121–12126.

Maines, T. R., Lu, X. H., Erb, S. M., Edwards, L., Guarner, J., Greer, P. W., Nguyen, D. C., Szretter, K. J., Chen, L. M., Thawatsupha, P., Chittaganpitch, M., Waicharoen, S., *et al.* (2005). Avian influenza (H5N1) viruses isolated from humans in Asia in 2004 exhibit increased virulence in mammals. *J. Virol.* **79**(18):11788–11800.

Massin, P., van der Werf, S., and Naffakh, N. (2001). Residue 627 of PB2 is a determinant of cold sensitivity in RNA replication of avian influenza viruses. *J. Virol.* **75**(11):5398–5404.

Matrosovich, M. N., Matrosovich, T. Y., Gray, T., Roberts, N. A., and Klenk, H. D. (2004). Human and avian influenza viruses target different cell types in cultures of human airway epithelium. *Proc. Natl. Acad. Sci. USA* **101**(13):4620–4624.

Matsuoka, Y., Swayne, D. E., Thomas, C., Rameix-Welti, M. A., Naffakh, N., Warnes, C., Altholtz, M., Donis, R., and Subbarao, K. (2009). Neuraminidase stalk length and additional glycosylation of the hemagglutinin influence the virulence of influenza H5N1 viruses for mice. *J. Virol.* **83**(9):4704–4708.

McAuley, J. L., Hornung, F., Boyd, K. L., Smith, A. M., McKeon, R., Bennink, J., Yewdell, J. W., and McCullers, J. A. (2007). Expression of the 1918 influenza A virus

PB1–F2 enhances the pathogenesis of viral and secondary bacterial pneumonia. *Cell Host Microbe* **2**(4):240–249.

Mitnaul, L. J., Matrosovich, M. N., Castrucci, M. R., Tuzikov, A. B., Bovin, N. V., Kobasa, D., and Kawaoka, Y. (2000). Balanced hemagglutinin and neuraminidase activities are critical for efficient replication of influenza A virus. *J. Virol.* **74**(13):6015–6020.

Mounts, A. W., Kwong, H., Izurieta, H. S., Ho, Y., Au, T., Lee, M., Buxton Bridges, C., Williams, S. W., Mak, K. H., Katz, J. M., Thompson, W. W., Cox, N. J., *et al.* (1999). Case-control study of risk factors for avian influenza A (H5N1) disease, Hong Kong, 1997. *J. Infect. Dis.* **180**(2):505–508.

Mubareka, S., Lowen, A. C., Steel, J., Coates, A. L., Garcia-Sastre, A., and Palese, P. (2009). Transmission of Influenza Virus via Aerosols and Fomites in the Guinea Pig Model. *J. Infect. Dis.* **199**(6):858–865.

Munster, V. J., de Wit, E., van Riel, D., Beyer, W. E., Rimmelzwaan, G. F., Osterhaus, A. D., Kuiken, T., and Fouchier, R. A. (2007). The Molecular Basis of the Pathogenicity of the Dutch Highly Pathogenic Human Influenza A H7N7 Viruses. *J. Infect. Dis.* **196**(2):258–265.

Muramoto, Y., Le, T. Q., Phuong, L. S., Nguyen, T., Nguyen, T. H., Sakai-Tagawa, Y., Horimoto, T., Kida, H., and Kawaoka, Y. (2006). Pathogenicity of H5N1 influenza A viruses isolated in Vietnam between late 2003 and 2005. *J. Vet. Med. Sci.* **68**(7):735–737.

Murphy, B. R., Hinshaw, V. S., Sly, D. L., London, W. T., Hosier, N. T., Wood, F. T., Webster, R. G., and Chanock, R. M. (1982). Virulence of avian influenza A viruses for squirrel monkeys. *Infect. Immun.* **37**(3):1119–1126.

Nakagawa, Y., Kimura, N., Toyoda, T., Mizumoto, K., Ishihama, A., Oda, K., and Nakada, S. (1995). The RNA polymerase PB2 subunit is not required for replication of the influenza virus genome but is involved in capped mRNA synthesis. *J. Virol.* **69**(2):728–733.

Nakajima, E., Morozumi, T., Tsukamoto, K., Watanabe, T., Plastow, G., and Mitsuhashi, T. (2007). A naturally occurring variant of porcine Mx1 associated with increased suscepti-bility to influenza virus *in vitro*. *Biochem. Genet.* **45**(1–2):11–24.

Nemeroff, M. E., Barabino, S. M., Li, Y., Keller, W., and Krug, R. M. (1998). Influenza virus NS1 protein interacts with the cellular 30 kDa subunit of CPSF and inhibits 3′end formation of cellular pre-mRNAs. *Mol. Cell.* **1**(7):991–1000.

Neumann, G., and Kawaoka, Y. (2002). Generation of influenza A virus from cloned cDNAs—historical perspective and outlook for the new millenium. *Rev. Med. Virol.* **12**(1):13–30.

Neumann, G., Watanabe, T., Ito, H., Watanabe, S., Goto, H., Gao, P., Hughes, M., Perez, D. R., Donis, R., Hoffmann, E., Hobom, G., and Kawaoka, Y. (1999). Generation of influenza A viruses entirely from cloned cDNAs. *Proc. Natl. Acad. Sci. USA* **96**(16):9345–9350.

Nguyen-Van-Tam, J. S., Nair, P., Acheson, P., Baker, A., Barker, M., Bracebridge, S., Croft, J., Ellis, J., Gelletlie, R., Gent, N., Ibbotson, S., Joseph, C., *et al.* (2006). Outbreak of low pathogenicity H7N3 avian influenza in UK, including associated case of human conjunc-tivitis. *Euro Surveill.* **11**(5):E060504.2.

Nicholls, J. M., Chan, M. C., Chan, W. Y., Wong, H. K., Cheung, C. Y., Kwong, D. L., Wong, M. P., Chui, W. H., Poon, L. L., Tsao, S. W., Guan, Y., and Peiris, J. S. (2007). Tropism of avian influenza A (H5N1) in the upper and lower respiratory tract. *Nat. Med.* **13**(2):147–149.

Ohuchi, M., Feldmann, A., Ohuchi, R., and Klenk, H. D. (1995). Neuraminidase is essential for fowl plague virus hemagglutinin to show hemagglutinating activity. *Virology* **212**(1):77–83.

Ohuchi, M., Ohuchi, R., Feldmann, A., and Klenk, H. D. (1997). Regulation of receptor binding affinity of influenza virus hemagglutinin by its carbohydrate moiety. *J. Virol.* **71**(11):8377–8384.

Olofsson, S., Kumlin, U., Dimock, K., and Arnberg, N. (2005). Avian influenza and sialic acid receptors: More than meets the eye? *Lancet Infect. Dis.* **5**(3):184–188.

Olsen, B., Munster, V. J., Wallensten, A., Waldenstrom, J., Osterhaus, A. D., and Fouchier, R. A. (2006). Global patterns of influenza a virus in wild birds. *Science* **312** (5772):384–388.

Olsen, S. J., Ungchusak, K., Sovann, L., Uyeki, T. M., Dowell, S. F., Cox, N. J., Aldis, W., and Chunsuttiwat, S. (2005). Family clustering of avian influenza A (H5N1). *Emerg. Infect. Dis.* **11**(11):1799–1801.

Opitz, B., Rejaibi, A., Dauber, B., Eckhard, J., Vinzing, M., Schmeck, B., Hippenstiel, S., Suttorp, N., and Wolff, T. (2007). IFNbeta induction by influenza A virus is mediated by RIG-I which is regulated by the viral NS1 protein. *Cell Microbiol.* **9**(4):930–938.

Pappas, C., Matsuoka, Y., Swayne, D. E., and Donis, R. O. (2007). Development and evaluation of an Influenza virus subtype H7N2 vaccine candidate for pandemic preparedness. *Clin. Vaccine Immunol.* **14**(11):1425–1432.

Park, C. H., Ishinaka, M., Takada, A., Kida, H., Kimura, T., Ochiai, K., and Umemura, T. (2002). The invasion routes of neurovirulent A/Hong Kong/483/97 (H5N1) influenza virus into the central nervous system after respiratory infection in mice. *Arch. Virol.* **147** (7):1425–1436.

Peiris, J. S., Yu, W. C., Leung, C. W., Cheung, C. Y., Ng, W. F., Nicholls, J. M., Ng, T. K., Chan, K. H., Lai, S. T., Lim, W. L., Yuen, K. Y., and Guan, Y. (2004). Re-emergence of fatal human influenza A subtype H5N1 disease. *Lancet* **363**(9409):617–619.

Peiris, M., Yuen, K. Y., Leung, C. W., Chan, K. H., Ip, P. L., Lai, R. W., Orr, W. K., and Shortridge, K. F. (1999). Human infection with influenza H9N2. *Lancet* **354**(9182):916–917.

Peper, R. L., and Van Campen, H. (1995). Tumor necrosis factor as a mediator of inflammation in influenza A viral pneumonia. *Microb. Pathog.* **19**(3):175–183.

Perdue, M. L., Garcia, M., Senne, D., and Fraire, M. (1997). Virulence-associated sequence duplication at the hemagglutinin cleavage site of avian influenza viruses. *Virus Res.* **49**(2):173–186.

Perrone, L. A., Plowden, J. K., Garcia-Sastre, A., Katz, J. M., and Tumpey, T. M. (2008). H5N1 and 1918 pandemic influenza virus infection results in early and excessive infiltration of macrophages and neutrophils in the lungs of mice. *PLoS Pathog.* **4**(8):e1000115.

Plotch, S. J., Bouloy, M., Ulmanen, I., and Krug, R. M. (1981). A unique cap(m7G pppXm)-dependent influenza virion endonuclease cleaves capped RNAs to generate the primers that initiate viral RNA transcription. *Cell* **23**(3):847–858.

Puzelli, S., Di Trani, L., Fabiani, C., Campitelli, L., De Marco, M. A., Capua, I., Aguilera, J. F., Zambon, M., and Donatelli, I. (2005). Serological analysis of serum samples from humans exposed to avian H7 influenza viruses in Italy between 1999 and 2003. *J. Infect. Dis.* **192**(8):1318–1322.

Rigoni, M., Shinya, K., Toffan, A., Milani, A., Bettini, F., Kawaoka, Y., Cattoli, G., and Capua, I. (2007). Pneumo- and neurotropism of avian origin Italian highly pathogenic avian influenza H7N1 isolates in experimentally infected mice. *Virology* **364**(1):28–35.

Rimmelzwaan, G. F., Kuiken, T., van Amerongen, G., Bestebroer, T. M., Fouchier, R. A., and Osterhaus, A. D. (2001). Pathogenesis of influenza A (H5N1) virus infection in a primate model. *J. Virol.* **75**(14):6687–6691.

Rimmelzwaan, G. F., van Riel, D., Baars, M., Bestebroer, T. M., van Amerongen, G., Fouchier, R. A., Osterhaus, A. D., and Kuiken, T. (2006). Influenza A virus (H5N1) infection in cats causes systemic disease with potential novel routes of virus spread within and between hosts. *Am. J. Pathol.* **168**(1):176–183quiz 364.

Rogers, G. N., and Paulson, J. C. (1983). Receptor determinants of human and animal influenza virus isolates: Differences in receptor specificity of the H3 hemagglutinin based on species of origin. *Virology* **127**(2):361–373.

Rohm, C., Zhou, N., Suss, J., Mackenzie, J., and Webster, R. G. (1996). Characterization of a novel influenza hemagglutinin, H15: Criteria for determination of influenza A subtypes. *Virology* **217**(2):508–516.

Rothwell, N. J. (1999). Annual review prize lecture cytokines—Killers in the brain? *J. Physiol.* **514**(Pt 1):3–17.

Rott, R., Orlich, M., and Scholtissek, C. (1979). Correlation of pathogenicity and gene constellation of influenza A viruses. III. Non-pathogenic recombinants derived from highly pathogenic parent strains. *J. Gen. Virol.* **44**(2):471–477.

Salomon, R., Franks, J., Govorkova, E. A., Ilyushina, N. A., Yen, H. L., Hulse-Post, D. J., Humberd, J., Trichet, M., Rehg, J. E., Webby, R. J., Webster, R. G., and Hoffmann, E. (2006). The polymerase complex genes contribute to the high virulence of the human H5N1 influenza virus isolate A/Vietnam/1203/04. *J. Exp. Med.* **203**(3):689–697.

Salomon, R., Hoffmann, E., and Webster, R. G. (2007a). Inhibition of the cytokine response does not protect against lethal H5N1 influenza infection. *Proc. Natl. Acad. Sci. USA* **104**(30):12479–12481.

Salomon, R., Staeheli, P., Kochs, G., Yen, H. L., Franks, J., Rehg, J. E., Webster, R. G., and Hoffmann, E. (2007b). *Mx1* gene protects mice against the highly lethal human H5N1 influenza virus. *Cell Cycle* **6**(19):2417–2421.

Scheiblauer, H., Kendal, A. P., and Rott, R. (1995). Pathogenicity of influenza A/Seal/Mass/1/80 virus mutants for mammalian species. *Arch. Virol.* **140**(2):341–348.

Scholtissek, C., Ludwig, S., and Fitch, W. M. (1993). Analysis of influenza A virus nucleoproteins for the assessment of molecular genetic mechanisms leading to new phylogenetic virus lineages. *Arch. Virol.* **131**(3–4):237–250.

Shinya, K., Ebina, M., Yamada, S., Ono, M., Kasai, N., and Kawaoka, Y. (2006). Avian flu: Influenza virus receptors in the human airway. *Nature* **440**(7083):435–436.

Shinya, K., Hamm, S., Hatta, M., Ito, H., Ito, T., and Kawaoka, Y. (2004). PB2 amino acid at position 627 affects replicative efficiency, but not cell tropism, of Hong Kong H5N1 influenza A viruses in mice. *Virology* **320**(2):258–266.

Shinya, K., Hatta, M., Yamada, S., Takada, A., Watanabe, S., Halfmann, P., Horimoto, T., Neumann, G., Kim, J. H., Lim, W., Guan, Y., Peiris, M., *et al.* (2005). Characterization of a human H5N1 influenza A virus isolated in 2003. *J. Virol.* **79**(15):9926–9932.

Skehel, J. J., and Wiley, D. C. (2000). Receptor binding and membrane fusion in virus entry: The influenza hemagglutinin. *Annu. Rev. Biochem.* **69**:531–569.

Smith, H., and Sweet, C. (1988). Lessons for human influenza from pathogenicity studies with ferrets. *Rev. Infect. Dis.* **10**(1):56–75.

Snyder, M. H., Buckler-White, A. J., London, W. T., Tierney, E. L., and Murphy, B. R. (1987). The avian influenza virus nucleoprotein gene and a specific constellation of avian and human virus polymerase genes each specify attenuation of avian-human influenza A/Pintail/79 reassortant viruses for monkeys. *J. Virol.* **61**(9):2857–2863.

Srinivasan, A., Viswanathan, K., Raman, R., Chandrasekaran, A., Raguram, S., Tumpey, T. M., Sasisekharan, V., and Sasisekharan, R. (2008). Quantitative biochemical rationale for differences in transmissibility of 1918 pandemic influenza A viruses. *Proc. Natl. Acad. Sci. USA* **105**(8):2800–2805.

Staeheli, P., Pitossi, F., and Pavlovic, J. (1993). Mx proteins: GTPases with antiviral activity. *Trends Cell Biol.* **3**(8):268–272.

Steel, J., Lowen, A. C., Mubareka, S., and Palese, P. (2009). Transmission of influenza virus in a mammalian host is increased by PB2 amino acids 627K or 627E/701N. *PLoS Pathog.* **5**(1):e1000252.

Steinhauer, D. A. (1999). Role of hemagglutinin cleavage for the pathogenicity of influenza virus. *Virology* **258**(1):1–20.

Stephenson, I., Nicholson, K. G., Wood, J. M., Zambon, M. C., and Katz, J. M. (2004). Confronting the avian influenza threat: Vaccine development for a potential pandemic. *Lancet Infect. Dis.* **4**(8):499–509.

Suarez, D. L., Senne, D. A., Banks, J., Brown, I. H., Essen, S. C., Lee, C. W., Manvell, R. J., Mathieu-Benson, C., Moreno, V., Pedersen, J. C., Panigrahy, B., Rojas, H., *et al.* (2004).

Recombination resulting in virulence shift in avian influenza outbreak, Chile. *Emerg. Infect. Dis.* **10**(4):693–699.

Subbarao, E. K., London, W., and Murphy, B. R. (1993). A single amino acid in the PB2 gene of influenza A virus is a determinant of host range. *J. Virol.* **67**(4):1761–1764.

Subbarao, K., Klimov, A., Katz, J., Regnery, H., Lim, W., Hall, H., Perdue, M., Swayne, D., Bender, C., Huang, J., Hemphill, M., Rowe, T., *et al.* (1998). Characterization of an avian influenza A (H5N1) virus isolated from a child with a fatal respiratory illness. *Science* **279**(5349):393–396.

Subbarao, K., and Luke, C. (2007). H5N1 viruses and vaccines. *PLoS Pathog.* **3**(3):e40.

Swayne, D. E., and Suarez, D. L. (2000). Highly pathogenic avian influenza. *Rev. Sci. Tech.* **19**(2):463–482.

Sweet, C., Bird, R. A., Cavanagh, D., Toms, G. L., Collie, M. H., and Smith, H. (1979). The local origin of the febrile response induced in ferrets during respiratory infection with a virulent influenza virus. *Br. J. Exp. Pathol.* **60**(3):300–308.

Szretter, K. J., Gangappa, S., Belser, J. A., Zeng, H., Chen, H., Matsuoka, Y., Sambhara, S., Swayne, D. E., Tumpey, T. M., and Katz, J. M. (2009). Early Control of H5N1 Influenza Virus Replication by the Type I Interferon Response in Mice. *J. Virol.* **83**(11):5825–5834.

Szretter, K. J., Gangappa, S., Lu, X., Smith, C., Shieh, W. J., Zaki, S. R., Sambhara, S., Tumpey, T. M., and Katz, J. M. (2007). Role of host cytokine responses in the pathogenesis of avian H5N1 influenza viruses in mice. *J. Virol.* **81**(6):2736–2744.

Tan, S. L., and Katze, M. G. (1998). Biochemical and genetic evidence for complex formation between the influenza A virus NS1 protein and the interferon-induced PKR protein kinase. *J. Interferon Cytokine Res.* **18**(9):757–766.

Tanaka, H., Park, C. H., Ninomiya, A., Ozaki, H., Takada, A., Umemura, T., and Kida, H. (2003). Neurotropism of the 1997 Hong Kong H5N1 influenza virus in mice. *Vet. Microbiol.* **95**(1–2):1–13.

Tannock, G. A., Paul, J. A., and Barry, R. D. (1985). Immunization against influenza by the ocular route. *Vaccine* **3**(Suppl. 3):277–280.

Taubenberger, J. K., Reid, A. H., Lourens, R. M., Wang, R., Jin, G., and Fanning, T. G. (2005). Characterization of the 1918 influenza virus polymerase genes. *Nature* **437**(7060):889–893.

Thompson, W. W., Shay, D. K., Weintraub, E., Brammer, L., Bridges, C. B., Cox, N. J., and Fukuda, K. (2004). Influenza-associated hospitalizations in the United States. *JAMA* **292**(11):1333–1340.

To, K. F., Chan, P. K., Chan, K. F., Lee, W. K., Lam, W. Y., Wong, K. F., Tang, N. L., Tsang, D. N., Sung, R. Y., Buckley, T. A., Tam, J. S., and Cheng, A. F. (2001). Pathology of fatal human infection associated with avian influenza A H5N1 virus. *J. Med. Virol.* **63**(3):242–246.

Tompkins, S. M., Lo, C. Y., Tumpey, T. M., and Epstein, S. L. (2004). Protection against lethal influenza virus challenge by RNA interference *in vivo*. *Proc. Natl. Acad. Sci. USA* **101**(23):8682–8686.

Tran, T. H., Nguyen, T. L., Nguyen, T. D., Luong, T. S., Pham, P. M., Nguyen, V. C., Pham, T. S., Vo, C. D., Le, T. Q., Ngo, T. T., Dao, B. K., Le, P. P., *et al.* (2004). Avian influenza A (H5N1) in 10 patients in Vietnam. *N. Engl. J. Med.* **350**(12):1179–1188.

Tumpey, T. M., Lu, X., Morken, T., Zaki, S. R., and Katz, J. M. (2000). Depletion of lymphocytes and diminished cytokine production in mice infected with a highly virulent influenza A (H5N1) virus isolated from humans. *J. Virol.* **74**(13):6105–6116.

Tumpey, T. M., Maines, T. R., Van Hoeven, N., Glaser, L., Solorzano, A., Pappas, C., Cox, N. J., Swayne, D. E., Palese, P., Katz, J. M., and Garcia-Sastre, A. (2007a). A two-amino acid change in the hemagglutinin of the 1918 influenza virus abolishes transmission. *Science* **315**(5812):655–659.

Tumpey, T. M., Suarez, D. L., Perkins, L. E., Senne, D. A., Lee, J. G., Lee, Y. J., Mo, I. P., Sung, H. W., and Swayne, D. E. (2002). Characterization of a highly pathogenic H5N1 avian influenza A virus isolated from duck meat. *J. Virol.* **76**(12):6344–6355.

Tumpey, T. M., Szretter, K. J., Van Hoeven, N., Katz, J. M., Kochs, G., Haller, O., Garcia-Sastre, A., and Staeheli, P. (2007b). The *Mx1* gene protects mice against the pandemic 1918 and highly lethal human H5N1 influenza viruses. *J. Virol.* **81**(19):10818–10821.

Tweed, S. A., Skowronski, D. M., David, S. T., Larder, A., Petric, M., Lees, W., Li, Y., Katz, J., Krajden, M., Tellier, R., Halpert, C., Hirst, M., *et al.* (2004). Human illness from avian influenza H7N3, British Columbia. *Emerg. Infect. Dis.* **10**(12):2196–2199.

Uiprasertkul, M., Puthavathana, P., Sangsiriwut, K., Pooruk, P., Srisook, K., Peiris, M., Nicholls, J. M., Chokephaibulkit, K., Vanprapar, N., and Auewarakul, P. (2005). Influenza A H5N1 replication sites in humans. *Emerg. Infect. Dis.* **11**(7):1036–1041.

Ungchusak, K., Auewarakul, P., Dowell, S. F., Kitphati, R., Auwanit, W., Puthavathana, P., Uiprasertkul, M., Boonnak, K., Pittayawonganon, C., Cox, N. J., Zaki, S. R., Thawatsupha, P., *et al.* (2005). Probable person-to-person transmission of avian influenza A (H5N1). *N. Engl. J. Med.* **352**(4):333–340.

Vacheron, F., Rudent, A., Perin, S., Labarre, C., Quero, A. M., and Guenounou, M. (1990). Production of interleukin 1 and tumour necrosis factor activities in bronchoalveolar washings following infection of mice by influenza virus. *J. Gen. Virol.* **71**(2):477–479.

Vahlenkamp, T. W., and Harder, T. C. (2006). Influenza virus infections in mammals. *Berl. Munch. Tierarztl. Wochenschr.* **119**(3–4):123–131.

Van Hoeven, N., Belser, J. A., Szretter, K. J., Zeng, H., Staeheli, P., Swayne, D. E., Katz, J. M., and Tumpey, T. M. (2009a). Pathogenesis of 1918 pandemic and H5N1 influenza virus infections in a guinea pig model: Antiviral potential of exogenous alpha interferon to reduce virus shedding. *J. Virol.* **83**(7):2851–2861.

Van Hoeven, N., Pappas, C., Belser, J. A., Maines, T. R., Zeng, H., Garcia-Sastre, A., Sasisekharan, R., Katz, J. M., and Tumpey, T. M. (2009b). Human HA and polymerase subunit PB2 proteins confer transmission of an avian influenza virus through the air. *Proc. Natl. Acad. Sci. USA* **106**(9):3366–3371.

Van Reeth, K. (2000). Cytokines in the pathogenesis of influenza. *Vet. Microbiol.* **74**(1–2):109–116.

van Riel, D., Munster, V. J., de Wit, E., Rimmelzwaan, G. F., Fouchier, R. A., Osterhaus, A. D., and Kuiken, T. (2006). H5N1 Virus Attachment to Lower Respiratory Tract. *Science* **312**(5772):399.

van Riel, D., Munster, V. J., de Wit, E., Rimmelzwaan, G. F., Fouchier, R. A., Osterhaus, A. D., and Kuiken, T. (2007). Human and avian influenza viruses target different cells in the lower respiratory tract of humans and other mammals. *Am. J. Pathol.* **171**(4):1215–1223.

Wagner, R., Matrosovich, M., and Klenk, H. D. (2002). Functional balance between haemagglutinin and neuraminidase in influenza virus infections. *Rev. Med. Virol.* **12**(3):159–166.

Wagner, R., Wolff, T., Herwig, A., Pleschka, S., and Klenk, H. D. (2000). Interdependence of hemagglutinin glycosylation and neuraminidase as regulators of influenza virus growth: A study by reverse genetics. *J. Virol.* **74**(14):6316–6323.

Walker, J. A., Molloy, S. S., Thomas, G., Sakaguchi, T., Yoshida, T., Chambers, T. M., and Kawaoka, Y. (1994). Sequence specificity of furin, a proprotein-processing endoprotease, for the hemagglutinin of a virulent avian influenza virus. *J. Virol.* **68**(2):1213–1218.

Wan, H., Sorrell, E. M., Song, H., Hossain, M. J., Ramirez-Nieto, G., Monne, I., Stevens, J., Cattoli, G., Capua, I., Chen, L. M., Donis, R. O., Busch, J., *et al.* (2008). Replication and transmission of H9N2 influenza viruses in ferrets: Evaluation of pandemic potential. *PLoS ONE* **3**(8):e2923.

Wang, X., Li, M., Zheng, H., Muster, T., Palese, P., Beg, A. A., and Garcia-Sastre, A. (2000). Influenza A virus NS1 protein prevents activation of NF-kappaB and induction of alpha/beta interferon. *J. Virol.* **74**(24):11566–11573.

Ward, A. C. (1997). Virulence of influenza A virus for mouse lung. *Virus Genes* **14**(3):187–194.

Webby, R. J., Perez, D. R., Coleman, J. S., Guan, Y., Knight, J. H., Govorkova, E. A., McClain-Moss, L. R., Peiris, J. S., Rehg, J. E., Tuomanen, E. I., and Webster, R. G. (2004).

Responsiveness to a pandemic alert: Use of reverse genetics for rapid development of influenza vaccines. *Lancet* **363**(9415):1099–1103.

Webster, R. G., Bean, W. J., Gorman, O. T., Chambers, T. M., and Kawaoka, Y. (1992). Evolution and ecology of influenza A viruses. *Microbiol. Rev.* **56**(1):152–179.

Webster, R. G., Geraci, J., Petursson, G., and Skirnisson, K. (1981). Conjunctivitis in human beings caused by influenza A virus of seals. *N. Engl. J. Med.* **304**(15):911.

Webster, R. G., Kawaoka, Y., and Bean, W. J., Jr. (1986). Molecular changes in A/Chicken/Pennsylvania/83 (H5N2) influenza virus associated with acquisition of virulence. *Virology* **149**(2):165–173.

Webster, R. G., Yakhno, M., Hinshaw, V. S., Bean, W. J., and Murti, K. G. (1978). Intestinal influenza: Replication and characterization of influenza viruses in ducks. *Virology* **84** (2):268–278.

World Health Organization Global Influenza Programme (2002). WHO manual on animal influenza diagnosis and surveillance. http://www.who.int/vaccine_research/diseases/influenza/WHO_manual_on_animal-diagnosis_and_surveillance_2002_5.pdf.

WHO (2009). Cumulative Number of Confirmed Human Cases of Avian Influenza A/(H5N1) Reported to WHO . W.H.O, Geneva.

Wyde, P. R., and Cate, T. R. (1978). Cellular changes in lungs of mice infected with influenza virus: Characterization of the cytotoxic responses. *Infect. Immun.* **22**(2):423–429.

Wyde, P. R., Peavy, D. L., and Cate, T. R. (1978). Morphological and cytochemical characterization of cells infiltrating mouse lungs after influenza infection. *Infect. Immun.* **21**(1):140–146.

Yao, L., Korteweg, C., Hsueh, W., and Gu, J. (2008). Avian influenza receptor expression in H5N1-infected and noninfected human tissues. *FASEB J.* **22**(3):733–740.

Yao, Y., Mingay, L. J., McCauley, J. W., and Barclay, W. S. (2001). Sequences in influenza A virus PB2 protein that determine productive infection for an avian influenza virus in mouse and human cell lines. *J. Virol.* **75**(11):5410–5415.

Yen, H. L., Aldridge, J. R., Boon, A. C., Ilyushina, N. A., Salomon, R., Hulse-Post, D. J., Marjuki, H., Franks, J., Boltz, D. A., Bush, D., Lipatov, A. S., Webby, R. J., *et al.* (2009). Changes in H5N1 influenza virus hemagglutinin receptor binding domain affect systemic spread. *Proc. Natl. Acad. Sci. USA* **106**(1):286–291.

Yen, H. L., Herlocher, L. M., Hoffmann, E., Matrosovich, M. N., Monto, A. S., Webster, R. G., and Govorkova, E. A. (2005a). Neuraminidase inhibitor-resistant influenza viruses may differ substantially in fitness and transmissibility. *Antimicrob. Agents Chemother.* **49**(10):4075–4084.

Yen, H. L., Lipatov, A. S., Ilyushina, N. A., Govorkova, E. A., Franks, J., Yilmaz, N., Douglas, A., Hay, A., Krauss, S., Rehg, J. E., Hoffmann, E., and Webster, R. G. (2007). Inefficient transmission of H5N1 influenza viruses in a ferret contact model. *J. Virol.* **81**(13):6890–6898.

Yen, H. L., Monto, A. S., Webster, R. G., and Govorkova, E. A. (2005b). Virulence may determine the necessary duration and dosage of oseltamivir treatment for highly pathogenic A/Vietnam/1203/04 influenza virus in mice. *J. Infect. Dis.* **192**(4):665–672.

Yuen, K. Y., Chan, P. K., Peiris, M., Tsang, D. N., Que, T. L., Shortridge, K. F., Cheung, P. T., To, W. K., Ho, E. T., Sung, R., and Cheng, A. F. (1998). Clinical features and rapid viral diagnosis of human disease associated with avian influenza A H5N1 virus. *Lancet* **351**(9101):467–471.

Zamarin, D., Garcia-Sastre, A., Xiao, X., Wang, R., and Palese, P. (2005). Influenza virus PB1–F2 protein induces cell death through mitochondrial ANT3 and VDAC1. *PLoS Pathog.* **1**(1):e4.

Zamarin, D., Ortigoza, M. B., and Palese, P. (2006). Influenza A virus PB1–F2 protein contributes to viral pathogenesis in mice. *J. Virol.* **80**(16):7976–7983.

Zell, R., Krumbholz, A., Eitner, A., Krieg, R., Halbhuber, K. J., and Wutzler, P. (2007). Prevalence of PB1–F2 of influenza A viruses. *J. Gen. Virol.* **88**(Pt 2):536–546.

Zitzow, L. A., Rowe, T., Morken, T., Shieh, W. J., Zaki, S., and Katz, J. M. (2002). Pathogenesis of avian influenza A (H5N1) viruses in ferrets. *J. Virol.* **76**(9):4420–4429.

CHAPTER 3

Virus Versus Host Cell Translation: Love and Hate Stories

**Anastassia V. Komarova,* Anne-Lise Haenni,[†]
and Bertha Cecilia Ramírez[‡]**

Contents

I.	Introduction	103
II.	Regulation Prior to Translation	104
	A. Editing	104
	B. Splicing	108
	C. Subgenomic RNA synthesis	110
III.	Initiation of Translation	111
	A. Cap-dependent initiation	111
	B. Closed-loop model or circularization	112
	C. VPg and initiation	115
	D. IRES-directed initiation	116
	E. Non-AUG initiation codons	119
	F. Multiple reading frames	122
	G. Modification of cell factors involved in initiation	128
IV.	Elongation of Translation	137
	A. Frameshift	137
	B. Modification of elongation factors	140
V.	Termination of Translation	141
	A. Readthrough	141
	B. Suppressor tRNAs	145
	C. Binding of release factors	146

* Viral Genomics and Vaccination Laboratory, Institut Pasteur, 25-28 Rue du Docteur Roux,
75015 Paris, France
† Institut Jacques Monod, CNRS, Universités Paris 6 et 7, 2 Place Jussieu, Tour 43, 75251 Paris, France
‡ Cellular Partners of Human Retroviruses, Institut Cochin, CNRS, Université Paris Descartes,
22 Rue Méchain, 75014 Paris, France, E-mail: cecilia.ramirez@inserm.fr

Advances in Virus Research, Volume 73
ISSN 0065-3527, DOI: 10.1016/S0065-3527(09)73003-9

VI. Conclusions   146
     Acknowledgments   147
     References   147

Abstract   Regulation of protein synthesis by viruses occurs at all levels of translation. Even prior to protein synthesis itself, the accessibility of the various open reading frames contained in the viral genome is precisely controlled. Eukaryotic viruses resort to a vast array of strategies to divert the translation machinery in their favor, in particular, at initiation of translation. These strategies are not only designed to circumvent strategies common to cell protein synthesis in eukaryotes, but as revealed more recently, they also aim at modifying or damaging cell factors, the virus having the capacity to multiply in the absence of these factors. In addition to unraveling mechanisms that may constitute new targets in view of controlling virus diseases, viruses constitute incomparably useful tools to gain in-depth knowledge on a multitude of cell pathways.

ABBREVIATIONS OF VIRUS NAMES

AAV-2	Adeno-associated virus type 2
AMCV	Artichoke mottled crinkle virus
APV	Achritosiphon pisum virus
ASLV	Avian sarcoma leukemia virus
BDV	Borna disease virus
BLV	Bovine leukemia virus
BNYVV	Beet necrotic yellow vein virus
BSBV	Beet soil-borne virus
BSMV	Barley stripe mosaic virus
BVQ	Beet virus Q
BWYV	Beet western yellows virus
BYDV	Barley yellow dwarf virus
BYV	Beet yellows virus
CaMV	Cauliflower mosaic virus
CarMV	Carnation mottle virus
CCFV	Cardamine chlorotic fleck virus
CCSV	Cucumber chlorotic spot virus
CoMV	Cocksfoot mottle virus
CNV	Cucumber necrosis virus
CPMV	Cowpea mosaic virus
CrPV	Cricket paralysis virus
CRSV	Carnation ringspot virus
CTV	Citrus tristeza virus

CVB	Coxsackie virus B
CyRSV	Cymbidium ringspot virus
DmeGypV	*Drosophila melanogaster* gypsy virus
EAV	Equine arterivirus
EBV	Epstein–Barr virus
EIAV	Equine infectious anemia virus
EMCV	Encephalomyocarditis virus
EqTV	Equine torovirus
FCV	Feline calicivirus
FMDV	Foot-and-mouth disease virus
HAstV	Human astrovirus
HAV	Hepatitis A virus
HBV	Hepatitis B virus
HCMV	Human cytomegalovirus
HCV	Hepatitis C virus
HDV	Hepatitis delta virus
HIV-1	Human immunodeficiency virus 1
HPIV-1	Human parainfluenza virus 1
HPV	Human papillomavirus
HRV	Human rhinovirus
HSV-1	Herpes simplex virus 1
HTLV-1	Human T-lymphotropic virus 1
IBV	Infectious bronchitis virus
LIYV	Lettuce infectious yellows virus
LRV1-1	Leishmania RNA virus 1-1
MCMV	Maize chlorotic mottle virus
MHV	Murine hepatitis virus
MLV	Murine leukemia virus
MMTV	Mouse mammary tumor virus
MNSV	Melon necrotic spot virus
MoMLV	Moloney murine leukaemia virus
NV	Norwalk virus
OCSV	Oat chlorotic stunt virus
PCMV	Peach chlorotic mottle virus
PCV	Peanut clump virus
PEBV	Pea early-browning virus
PEMV	Pea enation mosaic virus
PLRV	Potato leafroll virus
PPV	Plum pox virus
PSIV	Plautia stali intestine virus
PVM	Potato virus M
RCNMV	Red clover necrotic mottle virus
RhPV	Rhodopalosiphum padi virus

RTBV	Rice tungro bacilliform virus
SARS-CoV	Severe acute respiratory syndrome coronavirus
SbDV	Soybean dwarf virus
SBWMV	Soil-borne wheat mosaic virus
SceTy1V	Saccharomyces cerevisiae Ty1 virus
SceTy3V	Saccharomyces cerevisiae Ty3 virus
SCNMV	Sweet clover necrotic mottle virus
ScV-L-A	Saccharomyces cerevisiae virus L-A
SFV	Semliki Forest virus
SINV	Sindbis virus
STNV	Satellite tobacco necrosis virus
SV40	Simian virus 40
TBSV	Tomato bushy stunt virus
TCV	Turnip crinkle virus
TEV	Tobacco etch virus
TMEV	Theiler's murine encephalomyelitis virus
TMV	Tobacco mosaic virus
TNV	Tobacco necrosis virus
TRV	Tobacco rattle virus
TuMV	Turnip mosaic virus
VSV	Vesicular stomatitis virus
WDSV	Walleye dermal sarcoma virus

OTHER ABBREVIATIONS

aa	amino acid
CAT	chloramphenicol acetyltransferase
3′-CITE	3′-cap-independent translation element
CP	coat protein
eEF	eukaryotic elongation factor
eIF	eukaryotic initiation factor
eRF	eukaryotic release factor
4E-BP	eIF4E-binding protein
GCN2	general control nonderepressible-2
GP	glycoprotein
IGR	intergenic region
IRES	internal ribosome entry site
ITAF	IRES *trans*-acting factor
nt	nucleotide
ORF	open reading frame
P	phosphoprotein
PABP	poly(A) binding protein
Paip1	PABP-interacting protein 1

PCBP   poly(rC) binding protein
PERK   PKR-like endoplasmic reticulum kinase
PKR   protein kinase RNA
PTB   pyrimidine tract binding protein
RdRp   RNA-dependent RNA polymerase
sORF   short ORF
sg   subgenomic
TAV   transactivator
TC   ternary complex (eIF2-GTP-Met-tRNAi^{Met})
TE   translation enhancer
TLS   tRNA-like structure
unr   upstream of N-ras
uORF2   upstream ORF2
UTR   untranslated region
VPg   viral protein genome linked

I. INTRODUCTION

Because of the small size of their genomes and hence of their limited coding capacity, viruses have evolved a cohort of strategies to synthesize a few—and borrow from their host many—of the numerous elements required for their multiplication. The sophistication of the strategies elaborated by viruses is unsurpassed, and many of these strategies are common among viruses, but are rare or even nonexistent in uninfected cells. Many were first demonstrated in viral systems before being described in cell systems (reviewed in Bernardi and Haenni, 1998). The genome of viruses is compact and used to its limits: overlapping open reading frames (ORFs) are frequent, intergenic regions (IGRs) are usually short, and noncoding as well as coding regions are often involved in regulation of replication, transcription, and/or translation.

This chapter presents an overview of the strategies used by viruses of eukaryotes to regulate the expression of their viral genomes, ranging from the production of the RNA templates to translation of the encoded proteins. Emphasis is placed on RNA viruses, in which most of the strategies were originally described; moreover, only a few examples are taken from retroviruses, since the strategies used by these viruses have been discussed at length in several recent review articles (Balvay *et al.*, 2007; Brierley and Dos Ramos, 2006; Goff, 2004; Yilmaz *et al.*, 2006). For further information dealing with certain aspects of translation regulation mechanisms used by viruses, the reader may wish to turn to other reviews (Bushell and Sarnow, 2002; Gale *et al.*, 2000; Mohr *et al.*, 2007; Ryabova *et al.*, 2002).

II. REGULATION PRIOR TO TRANSLATION

Viruses use several regulation strategies prior to translation to obtain maximum protein diversity from their small genomes. Prior to initiation of translation, the viral RNA, due to serve as template for protein synthesis, can be modified so as to favor synthesis of certain viral proteins, sometimes to the detriment of cell proteins. This can be achieved by various mechanisms such as editing, splicing, and the production of subgenomic (sg) RNAs including cap-snatching. The importance of regulation at this level has, moreover, been highlighted in recent publications showing that viral translation and transcription are coupled (Barr, 2007; Katsafanas and Moss, 2007; Sanz *et al.*, 2007).

A. Editing

Editing is a mechanism in which an RNA-encoded nucleotide (nt) is modified, or one, two, or more pseudotemplated nts are inserted at the editing site; various forms of editing have been described (Weissmann *et al.*, 1990). Viruses resort to editing by nt modification in the case of Hepatitis delta virus (HDV; genus Deltavirus), and by the addition of one or more nts in paramyxoviruses.

1. Editing by nucleotide modification

HDV is a highly pathogenic subviral particle totally dependent on the DNA virus Hepatitis B virus (HBV; family Hepadnaviridae) for its propagation (reviewed in Taylor, 2006); it requires the HBV envelope proteins to assemble into HDV particles. The genome of HDV is a (−) sense, closed circular, and highly structured single-stranded RNA (of ∼ 1680 nts in HDV genotype III) referred to as the genomic RNA (Fig. 1); it is devoid of coding capacity (i.e., devoid of ORF). However, the complementary antigenomic RNA contains a unique ORF for the short surface antigen HDAg-S of 195 amino acids (aa); HDAg-S is produced from an 800-nt long linear sg mRNA that is both capped and polyadenylated (Gudima *et al.*, 2000). The protein is produced throughout infection and is required for HDV replication. At late times in infection, editing of the antigenomic RNA occurs by deamination of the A residue (position 1012) of the UAG codon that ends the HDAg-S ORF; editing does not occur on the HDAg-S mRNA. Hence the antigenome, which is the template for editing, must be replicated to yield the edited genomic RNA prior to being transcribed to produce the edited sg mRNA that is also capped and polyadenylated. Editing also requires previous refolding of the antigenome, from a rod-like to a branched double-hairpin structure in HDV genotype III, or to a highly conserved base-paired structure in HDV genotype I (Casey, 2002; Cheng *et al.*, 2003).

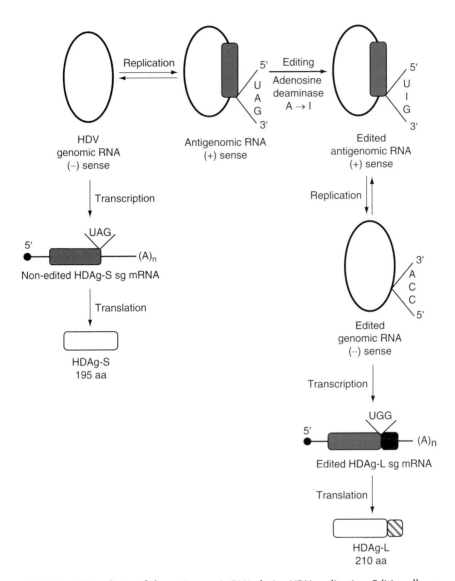

FIGURE 1   RNA editing of the antigenomic RNA during HDV replication. Editing allows the virus to express two proteins HDAg-S and HDAg-L from one coding sequence. The black circle represents the 5′-capped terminus of the mRNA. (A)$_n$ represents a poly(A) tail. Gray rectangles represent ORFs, the black rectangle represents the edited ORF region, the white rectangles represent the HDAg protein, and the gray slashed rectangle represents the extended C-terminal region in the edited protein.

   Deamination of the A residue in UAG leading to an I (inosine) residue and producing the triplet UIG (Fig. 1) is triggered by a host adenosine deaminase that acts on RNA substrates. Upon replication of the edited antigenome, the I residue recognized as G leads to ACC in the edited genomic RNA that is then transcribed as UGG coding for tryptophane in the edited mRNA. As a consequence, the edited sg mRNA presents an extended ORF and produces HDAg-L of 210 aa. The two viral proteins share the same N-terminal region, the longer protein bearing an extended C-terminal region; they are responsible for two distinct functions in the HDV-infected cell. The longer protein inhibits replication and editing and is necessary for virus assembly, whereas the shorter protein is required for replication (Cheng *et al.*, 2003). Editing is, therefore, a vital process for HDV propagation, and an exquisite balance between the nonedited and edited mRNAs, and between replication and virus production is a major factor in maintaining optimum virus production. How this equilibrium is reached remains largely speculative, although editing is known to involve specific structural elements that depend on the HDV genotype considered (Casey, 2002; Cheng *et al.*, 2003).

2. Editing by nucleotide addition

In a coding RNA, the introduction of nontemplated nts leads to the production of a new edited mRNA. In such an mRNA, a change in reading frame at the point of editing has occurred, resulting in the synthesis of a new protein. The new "edited" protein is identical to the "original" protein resulting from the nonedited mRNA, from the 5′ terminus to the editing site, but different thereafter. The protein resulting from editing is usually endowed with properties and/or activities that are absent from the original protein.

Paramyxoviruses are animal viruses that belong to the order Mononegavirales. They possess a nonsegmented (also known as monopartite) (−) strand RNA genome of 15–16 kb (reviewed in Nagai, 1999). Their genome encodes a minimum of six structural proteins that are produced from six capped and polyadenylated mRNAs. In the complementary antigenomic RNA, the ORFs are separated by conserved sequences that dictate initiation and termination of the six transcripts. Except for the phosphoprotein (P) mRNA, each mRNA expresses a single protein from a single ORF. The P gene is more complex. In most members of the subfamily Paramyxovirinae (family Paramyxoviridae), editing of the P mRNA results in the insertion of 1–5 nontemplated G residues within a run of Gs at the level of a conserved A_nG_n editing sequence (Cattaneo *et al.*, 1989; Mahapatra *et al.*, 2003; Steward *et al.*, 1993; reviewed in Strauss and Strauss, 1991), presumably as a result of a stuttering process (Vidal *et al.*, 1990). This causes a shift within the P ORF and may lead to the synthesis of up to six nonstructural proteins from edited and nonedited mRNAs depending on the virus. In Sendai virus, two mRNAs can be produced by editing of the P/C (also known as P) mRNA, the V mRNA (insertion of 1 G), and the W mRNA (insertion of 2 or 5 Gs) (Fig. 2A)

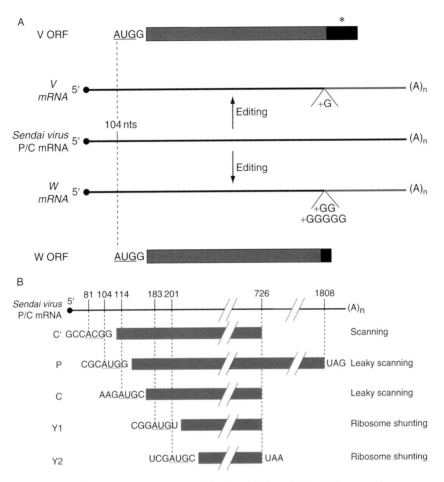

FIGURE 2 Schematic representation of the Sendai virus P/C mRNA expression. (A) Regulation prior to translation: editing of P/C mRNA. *, Cysteine-rich coding region. Initiation codon is underlined. (B) Regulation at initiation of translation of P/C mRNA. Initiation codons of each ORF are underlined and the sequence contexts are presented; termination codons are shown. Other indications are as in legend of Fig. 1.

(Curran *et al.*, 1991). The V protein, a cysteine-rich protein, binds Zn^{+2}, a characteristic related to virus pathogenicity in mice. Indeed, mutation of the cysteine residues in the corresponding V protein in the Sendai virus genome reduces Zn^{+2} binding and pathogenicity (Fukuhara *et al.*, 2002).

Editing is also observed for the synthesis of the structural glycoproteins (GPs) of Ebola virus (family Filoviridae, order Mononegavirales), whose monopartite (−) strand RNA genome contains seven genes. Two GPs are produced from the GP gene, a short and a long form that make up

80% and 20% of the total GP protein synthesized, respectively; they differ in their C-terminal region. The short form of GP is produced by the unedited transcript, whereas the long form results from an edited transcript that has acquired an additional nontemplated A residue within a stretch of seven conserved A residues in the GP ORF. The long form possesses a transmembrane anchor sequence absent from the short form (Sanchez *et al.*, 1996; Volchkov *et al.*, 1995).

B. Splicing

Splicing is a strategy used by DNA viruses such as those of the family Adenoviridae and Polyomaviridae (reviewed in Ziff, 1980, 1985), the Caulimoviridae (reviewed in Ryabova *et al.*, 2006), the Baculoviridae (Chisholm and Henner, 1988; Kovacs *et al.*, 1991), and of the genus Mastrevirus, family Geminiviridae (Schalk *et al.*, 1989). It is less frequently encountered among RNA viruses, although it is observed in certain RNA viruses that replicate in the nucleus. This is the case of retroviruses whose mRNAs undergo a complicated cascade of splicing and alternative splicing events. The splicing mechanisms used by these viruses will not be developed here, having received considerable attention in several review articles (Cullen, 1998; Stoltzfus and Madsen, 2006). Examples of nonretroviruses whose RNA genomes multiply in the nucleus and employ splicing are briefly presented here; they are Borna disease virus (BDV) and Influenza virus.

BDV (family Bornaviridae) belongs to the order Mononegavirales. However, it differs from the other members of this order by several unique features (reviewed in de la Torre, 2002; Tomonaga *et al.*, 2002). As opposed to the other members of this order whose life cycle occurs entirely in the cytoplasm, BDV is replicated and transcribed in the nucleus of the infected cell and employs the cellular RNA splicing machinery. The two splice donor and three splice acceptor sites follow the general mammalian splice site consensus (Fig. 3). The six ORFs contained in the antigenome are not separated by conserved IGRs as in other mononegavirales. Rather, the six proteins of BDV are translated from capped and polyadenylated transcripts that are initiated at only three sites (S1–S3) and terminate at five possible sites (T1–T4, t6). The nucleoprotein (known as N) is produced from a transcript initiated at S1, and the X protein and P from a transcript initiated at S2. The matrix (M), glycoprotein (G), and polymerase (L) are all produced from transcripts initiated at S3, and resort to alternative splicing for the production of the transcripts required. Splicing of intron I that overlaps the M ORF abolishes synthesis of the corresponding protein and produces protein G, while splicing of introns I and II (the latter corresponds to most of the G ORF) leads to the synthesis of the L protein. Additionally,

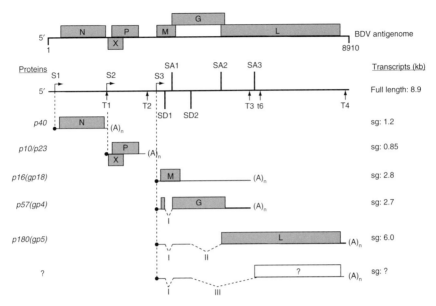

FIGURE 3 Regulation prior to translation by splicing in BDV. S1–S3, transcription initiation sites; T1–T4 and t6, polyadenylation/termination sites; SA1–SA3, splice acceptor sites; SD1 and SD2, splice donor sites. I, II, and III are introns. N, nucleoprotein; P, phosphoprotein; M, matrix; G, glycoprotein; and L, polymerase. The open rectangle corresponds to the ORF of a putative protein. Other indications are as in legend of Fig. 1.

splicing of intron III that uses the same 5' splice donor site as intron II but another 3' splice acceptor site also eliminates most of the G ORF as well as the 5' region of the L ORF. This could lead to the production of yet another BDV protein; this putative protein has so far not been identified (reviewed in Jordan and Lipkin, 2001). Although translation of M is prevented by splicing of intron I, this leaves a minicistron corresponding to the N-terminal region of M which enhances translation of G, presumably by promoting ribosomal reinitiation. However, mutation experiments using the unspliced transcript, suggest that leaky scanning is also a mechanism that could lead to the synthesis of G from the unspliced transcript (Schneider *et al.*, 1997).

Influenza viruses (family Orthomyxoviridae) are enveloped viruses with a segmented (−) strand RNA genome; they are replicated and transcribed in the nucleus by the viral RNA-dependent RNA polymerase (RdRp) complex composed of PB1, PB2, and PA. In the nucleus of infected cells, transcription of the viral RNAs into mRNAs by the RdRp requires cooperation with ongoing transcription by the cellular RNA polymerase II, since the RdRp initiates synthesis of viral mRNAs via cap-snatching using capped cellular mRNAs (see below; reviewed in

Lamb and Krug, 2001; Rao *et al.*, 2003). Influenza A virus and Influenza B virus are composed of eight RNA segments. Alternative splicing leads to the synthesis of two proteins from segments seven and eight of Influenza A virus. Regulation of the choice of the 5′ or 3′ splice sites is finely controlled. Although alternative splicing occurs in many viruses, only in a few cases have viral proteins been shown to be involved in this mechanism. In segment seven of Influenza A virus, two alternative 5′ splice sites control the production of the shorter (mRNA$_3$: 111 nts) and the longer (M2 mRNA: 151 nts) spliced mRNAs from the pre-mRNA known as M1 mRNA. Both spliced RNAs use the same 3′ splice site. At early times after infection, the more favorable upstream 5′ splice site is used, leading to the synthesis of mRNA$_3$ that potentially codes for a 9-aa peptide (as yet undetected). At later times after infection, the RdRp complex now produced in sufficient amounts binds to and blocks the upstream 5′ splice site, forcing the cell splicing machinery to switch to the less favorable downstream 5′ splice site. As a consequence, M2 mRNA is synthesized as is also its encoded M2 ion channel protein of 97 aa (Shih *et al.*, 1995).

C. Subgenomic RNA synthesis

Contrary to mRNAs of eukaryotic cells that are largely monocistronic, the RNA genomes of many eukaryotic viruses contain multiple ORFs of which generally only the 5′-proximal ORF is accessible for translation. Thus, viruses have evolved several strategies to synthesize the proteins corresponding to 5′-distal ORFs (reviewed in Miller and Koev, 2000; White, 2002). One of the most common mechanisms is the production of 3′-coterminal sgRNAs. In such templates, the internally positioned and the 3′-proximal ORFs in the genome of (+) strand RNA viruses are accessed by sgRNAs in which these ORFs become 5′-proximal and serve as mRNAs. sgRNAs are generally synthesized by internal initiation of RNA synthesis on the complementary (−) RNA strand. They are 5′-truncated versions of the genomic RNA and therefore perfect copies of the region of the genome from which they derive.

A particular mechanism of sgRNA production is used by RNA viruses whose genome segments are ambisense or of (−) polarity and resort to cap-snatching. This mechanism was first described for the synthesis of the mRNAs of *Influenza virus* (Bouloy *et al.*, 1978; Krug *et al.*, 1979). The endonuclease activity of the viral RdRp cleaves nuclear cellular capped RNAs to generate capped primers of up to about 20 nts in length for viral mRNA synthesis. As a result, the viral mRNAs contain capped nonviral oligonucleotides at their 5′ end. Several plant (members of the family *Bunyaviridae* and of the genus *Tenuivirus*) and animal (members of the family *Bunyaviridae*) viruses with (−) strand or ambisense RNA genomes also use this transcription initiation mechanism. Since these viruses multiply in the cytoplasm, they use cytoplasmic rather than nuclear cellular capped RNAs as primers (Garcin and Kolakofsky, 1990; Garcin *et al.*, 1995; Huiet *et al.*, 1993; Raju *et al.*, 1990; Ramírez *et al.*, 1995; Vialat and Bouloy, 1992).

III. INITIATION OF TRANSLATION

A. Cap-dependent initiation

The most common strategy of translation initiation encountered among eukaryotes is cap-dependent translation (reviewed in Jackson and Kaminski, 1995; Pestova *et al.*, 2007). This occurs in capped, generally monocistronic mRNAs, whose initiation codon lies close to the 5' cap structure, and whose leader sequence also called 5' untranslated region (UTR) possesses varying degrees of secondary structure. A number of complex steps lead to binding of the small 40S ribosomal subunit to the mRNA. The assembly of the eukaryotic initiation factor (eIF) 2, GTP, and Met-tRNAi^{Met} forms the ternary complex (TC). Interaction of the TC with the 40S ribosomal subunit, facilitated by eIF1, eIF1A, and eIF3, leads to the formation of the 43S preinitiation complex. eIF3 is composed of 13 subunits (eIF3a–eIF3m) (Hinnebusch, 2006). The cap structure is recognized by the heterotrimer eIF4F composed of eIF4G (multivalent scaffolding protein), eIF4E (cap-binding protein), and eIF4A (ATP-dependent helicase). The 43S preinitiation complex binds to the 5' end of the mRNA with the help of eIF4F in the presence of eIF4B, and the complex scans the mRNA leader sequence until it reaches the initiation codon to form the 48S initiation complex (Kozak and Shatkin, 1978). The initiation codon is usually the first AUG codon encountered; it is recognized by base-pairing with the anticodon of Met-tRNAi^{Met} and the efficiency of recognition depends on the sequence context surrounding the initiation codon. The most favorable context in mammals is RCC<u>A</u>UGG with purine (R) at position -3 (Kozak, 1986, 1991), and in plants it is ACA<u>A</u>UGG (Fütterer and Hohn, 1996). At this step the 48S initiation complex is joined by the large 60S ribosomal subunit to form the 80S ribosome. Joining requires two additional factors: eIF5 and eIF5B. Hydrolysis of eIF2-bound GTP induced by eIF5 leads to reduction in the affinity of eIF2 for Met-tRNAi^{Met}. In turn, the essential ribosome-dependent GTPase activity of eIF5B leads to displacement of the eIF2-bound GDP and other initiation factors from the 40S subunit (reviewed in Pestova *et al.*, 2007). The assembled 80S ribosome contains the initiator Met-tRNAi^{Met} in the ribosomal P (peptidyl) site and another aa-tRNA in the ribosomal A (aminoacyl) site. The delivery of the aa-tRNA is mediated by the eukaryotic elongation factor (eEF) 1A–GTP complex. After peptide bond formation (triggered by the peptidyl transferase in the ribosome) eEF2 binding and subsequent GTP hydrolysis catalyze ribosomal translocation, and the elongation cycle begins (Frank *et al.*, 2007). This general strategy is also adopted by a large number of eukaryotic viruses. Yet the RNA genome of certain viruses lacks a cap structure; the 5′ end of such RNA genomes can carry a covalently bound viral protein designated viral protein genome-linked (VPg), or begin with a di- (or a mono-) phosphate. In other cases, the 5′ UTR of the viral RNA contains an internal ribosome entry site (IRES) responsible for initiation of translation.

B. Closed-loop model or circularization

In most eukaryotic mRNAs, the 5′ cap structure and the 3′ poly(A) tail appear to work together leading to efficient translation initiation. This is believed to occur when the 5′ and 3′ ends are brought in close proximity, referred to as the mRNA circularization or closed-loop model. The existence of cellular polyribosomes arranged in a circle was visualized using electron microscopy (Christensen *et al.*, 1987). Circularization is brought about by binding of the initiation factor eIF4E to the 5′ cap and to eIF4G. In turn, eIF4G binds to the poly(A)-binding protein (PABP) bound to the 3′ poly(A) tail (Fig. 4A). PABP contains four conserved RNA recognition motifs in its N-terminal domain that are involved in RNA and eIF4G interactions, and a C-terminal domain that binds to several proteins, including eIF4B, the eukaryotic release factor (eRF) 3 and the PABP-interacting protein 1 (Paip1). Therefore, PABP promotes the formation of the closed-loop complex by binding directly to eIF4G (Gale *et al.*, 2000; Gallie, 1998; Imataka *et al.*, 1998; reviewed in Dreher and Miller, 2006) or through Paip1 binding to eIF4A (Craig *et al.*, 1998) or by PABP interaction with eIF4B (Bushell *et al.*, 2001; Le *et al.*, 1997) (Fig. 4A). Circularization thus appears to be mediated by RNA–protein and protein–protein interactions. Increasing evidence has been provided for the involvement of both 5′ and 3′ UTRs of eukaryotic mRNAs and viral mRNAs in initiation of translation (reviewed in Edgil and Harris, 2006; Hentze *et al.*, 2007; Komarova *et al.*, 2006; Mazumder *et al.*, 2003; Wilkie *et al.*, 2003). In addition to contributing to mRNA stabilization, circularization probably facilitates ribosome recruiting from the 3′ end of the mRNA after a terminated round of translation, to the 5′ region for initiation of a second round. This could be achieved via interaction of PABP with eRF3 that by interacting with eRF1, would result in the formation of a closed loop by way of the 5′ cap—eIF4E–eIF4G–PABP–eRF3–eRF1—termination codon (Fig. 4A) (Uchida *et al.*, 2002).

Certain viral mRNAs are devoid of 5′ cap structure or VPg, and some of them are devoid of 3′ poly(A) tail. Nevertheless, such mRNAs are highly efficient, and circularization presumably required for efficient

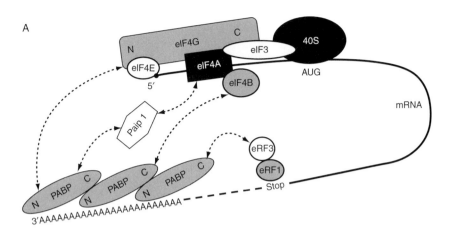

A

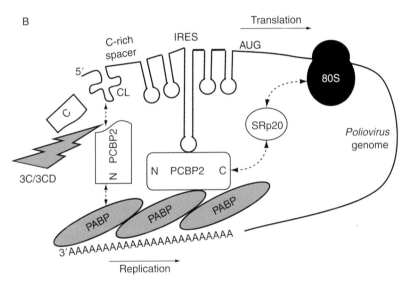

B

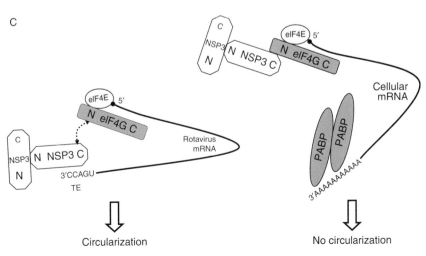

C

translation is achieved via various mechanisms. In the case of viral RNAs with a poly(A) tail, circularization has been investigated by looking for viral and/or host proteins that participate in circularization. For instance, in Poliovirus (family Picornaviridae) RNA whose VPg is removed soon after entry of the virus into the cell, the long 5' UTR of its genomic RNA contains an IRES preceded by a cloverleaf structure (Fig. 4B). PABP interacts with the poly(A) tail of the viral RNA. It also binds to the poly (rC)-binding protein 2 (PCBP2) that binds to the IRES structure to circularize the viral mRNA for translation (Blyn *et al.*, 1997; Silvera *et al.*, 1999; Walter *et al.*, 2002). Binding of PCBP2 to PABP leads to circularization of the viral RNA with the formation of an RNA–protein–protein–RNA bridge. In addition, the cellular protein SRp20 that is involved in cellular mRNA splicing and nucleocytoplasmic trafficking and also cofractionates with ribosomal subunits, interacts with PCBP2, and promotes *Poliovirus* IRES-driven translation (Bedard *et al.*, 2007).

Circularization can also be achieved by direct base pairing between a region in the 5' UTR and a region in the 3' UTR of an mRNA; it is used in particular by viral mRNAs that possess neither cap (or VPg) nor poly(A) tail. In several instances, specific interactions have been detected and their functional significance investigated by phylogenetic studies of conserved regions within the ends of viral RNAs, and by mutation analyses of the base-paired regions presumably involved (reviewed in Miller and White, 2006). Among plant RNA viruses, the region of the 3' UTR required for translation is frequently referred to as 3'-cap-independent translation element (3'-CITE). Several classes of 3'-CITEs have been described (reviewed in Miller *et al.*, 2007). They presumably operate by long-distance base pairing between the 3' UTR and a complementary region in the 5' UTR, leading to interactions known as kissing stem–loop interactions. Some of the well-studied cases are those of Tobacco necrosis virus (TNV; family Tombusviridae; Meulewaeter *et al.*, 2004; Shen and Miller, 2004), Satellite tobacco necrosis virus (STNV; Guo *et al.*, 2001; Meulewaeter *et al.*, 1998) Tomato bushy stunt virus (TBSV; family Tombusviridae; Fabian and White, 2004, 2006), Barley yellow dwarf virus (BYDV; family Luteoviridae; Guo et al., 2000, 2001), and Maize necrotic streak virus (family Tombusviridae; Scheets and Redinbaugh, 2006).

FIGURE 4   Possible models of circularization. (A) Closed-loop model or circularization of cellular mRNAs. eIF4E+eIF4G+eIF4A, eIF4F complex; Stop, termination codon. (B) Models of circularization of Poliovirus genome. CL, cloverleaf structure; 3C and 3CD, viral proteases. (C) Role of rotavirus NSP3 in mRNA circularization. NSP3 mediates viral mRNA circularization (left) and inhibits cellular mRNA circularization (right). N and C, N- and C-terminal regions of proteins; TE, translation enhancer. Dashed arrows indicate interactions between proteins. Other indications are as in legend of Fig. 1.

An interesting outcome of these studies has been the observation that in certain cases, part of the $3'$ UTR can function by recruiting initiation factors required for initiation of translation. This seems to be the case of STNV in which a $3'$ translation enhancer (TE) mimics a $5'$ cap structure (Gazo *et al.*, 2004) by binding to eIF4E and this binding is enhanced by eIF4G (or eIFiso4E and eIFiso4G of eIFiso4F, isoforms only found in plants). Moreover, the $3'$ UTR of STNV RNA contains a region that has been reported to be complementary of the $3'$ end of the ribosomal 18S RNA (Danthinne *et al.*, 1993). Thus, long-distance RNA–RNA interaction between the $5'$ and $3'$ UTRs might bring eIF4F and the 40S subunit positioned on $3'$ UTR close to the initiation codon, favoring initiation of translation. Phylogenetic studies suggest that similar mechanisms may be involved in stimulating translation of other viral mRNAs devoid of cap and poly(A) tail (Gazo *et al.*, 2004; Miller *et al.*, 2007; Shen and Miller, 2004; Treder *et al.*, 2008).

Rotaviruses (family Reoviridae) contain 11 double-stranded generally monocistronic RNAs; they are capped, and most of them contain a short conserved sequence (UGACC) at their $3'$ end. This sequence serves as $3'$ TE. Enhanced gene expression by the $3'$ TE requires the viral nonstructural protein NSP3 (Fig. 4C). This protein binds not only to the $3'$ TE but also to eIF4G, suggesting that it behaves as a functional homolog of PABP, leading to circularization of the mRNA (Piron *et al.*, 1998). Moreover, upstream of the common UGACC sequence in its $3'$ UTR, the mRNA of gene 6 coding for the structural protein VP6 possesses a unique gene-specific TE that does not require NSP3 for activity (Yang *et al.*, 2004).

Finally, in the tripartite (+) strand RNA virus Alfalfa mosaic virus (family Bromoviridae), whose RNAs are capped but lack a poly(A) tail, a few molecules of coat (also known as capsid) protein (CP) appear to replace PABP in promoting translation: the CP binds strongly to specific regions in the $3'$ UTR and also to eIF4G (or eIFiso4G; Krab *et al.*, 2005; reviewed in Bol, 2005). When an artificial poly(A) tail is tagged to the $3'$ UTR, CP molecules are no longer required for translation (Neeleman *et al.*, 2001).

C. VPg and initiation

The presence of a VPg linked covalently to the $5'$ end of an RNA is characteristic of members of various virus families such as the Birnaviridae, Caliciviridae, Picornaviridae, Potyviridae, Comoviridae, and Luteoviridae (reviewed in Sadowy *et al.*, 2001). The size of the VPg varies from 3 kDa (members of the Picornaviridae family) to 90 kDa (members of the Birnaviridae family). Binding of VPg to eIF3 and eIF4E suggests that an initiation complex is formed and recruited to the viral mRNA, a complex in which VPg would behave as a cap substitute. VPg may therefore interfere with translation by interacting with initiation factors that are required for initiation of both cap-dependent and IRES-containing mRNA translation.

In Poliovirus, whose genome contains a VPg and an IRES in its 5′ UTR, the VPg is removed from the genomic RNA early in infection and the viral mRNA lacks a VPg. Therefore, VPg does not regulate initiation of translation in this virus and probably in other members of the *Picornaviridae* family.

The genome of Turnip mosaic virus (TuMV, family Potyviridae) is devoid of IRES and cap structure. Its VPg in the precursor form 6K-VPg-Pro appears to favor translation of viral proteins by interacting with eIFiso4E (Leonard *et al.*, 2004; Wittmann *et al.*, 1997). TuMV and Tobacco etch virus (TEV; family, Potyviridae) can interfere *in vitro* with the formation of a translation initiation complex on host plant cellular mRNA by sequestering eIFiso4E, since the binding affinity of VPg for eIFiso4E is stronger than that of capped RNA. VPg enhances uncapped viral mRNA translation and inhibits capped mRNA translation. Moreover, it appears to function as an alternative cap-like structure by forming a complex with eIFiso4E and eIFiso4G (Khan *et al.*, 2008; Miyoshi *et al.*, 2006).

Furthermore, for viruses such as those of the family Caliciviridae that are also devoid of cap or IRES, evidence for the involvement of the VPg in translation initiation has been documented: in addition to binding to the eIF3 complex (in particular to its eIF3d subunit), the Norwalk virus (NV) VPg inhibits translation of cap-dependent and of IRES-containing reporter mRNAs *in vitro* (Daughenbaugh *et al.*, 2003). In Feline calicivirus (FCV), the VPg directly interacts with eIF4E *in vitro* (Goodfellow *et al.*, 2005) and removal of the VPg from the FCV RNA results in dramatic reduction of viral protein synthesis (Herbert *et al.*, 1997).

D. IRES-directed initiation

Initiation of translation of the genome of numerous RNA viruses does not comply with the general cap-dependent scanning mechanism of eukaryotic protein synthesis (reviewed in Doudna and Sarnow, 2007; Kneller *et al.*, 2006). Rather, initiation can occur downstream of a (usually) long GC-rich 5′ UTR known as IRES that in contrast to classical cap-dependent initiation of translation, plays an active role in 40S ribosomal subunit recruitment. These viral 5′ UTRs are generally highly structured, thereby hindering movement of the scanning ribosomes. Animal viruses that resort to this strategy are picornaviruses (reviewed in Belsham and Jackson, 2000; Martínez-Salas and Fernández-Miragall, 2004; Martínez-Salas *et al.*, 2001; Pestova *et al.*, 2001) and pestiviruses (Pisarev *et al.*, 2005).

Ribosomal entry directly to an internal AUG initiation codon on an mRNA devoid of cap structure was demonstrated by placing an IRES between two mRNA cistrons in a dicistronic construct. The presence of the IRES allowed the expression of the downstream cistron independently of the upstream cistron (Jang *et al.*, 1988; Pelletier and Sonenberg, 1988). Hence, *cis*-acting elements in the IRES appear to cap independently recruit ribosomes to the initiation codon, and the 5′ UTR can be considered an IRES if it drives initiation of translation of the downstream cistron.

Considerable work has been directed toward deciphering the sequence elements involved in IRES-mediated initiation, and the protein factors participating in this step of translation. Translation by an IRES obviates the need of certain host eIFs (that differ for different groups of IRESs), and often requires additional host proteins, the IRES *trans*-acting factors (ITAFs). These are mRNA-binding proteins such as the pyrimidine tract-binding protein (PTB), ITAF45, PCBP2, the cellular cytoplasmic RNA-binding protein designated upstream of N-ras (unr), and the La autoantigen. The IRESs involved in initiation are part of the 5′ UTR or of the IGR, and their integrity is required for full activity. They sometimes include 5′ nts of the ORF following the IRES (Rijnbrand *et al.*, 2001). Based on their sequence and structure, the IRESs of members of the Picornaviridae family can be divided into three major groups; (1) Enterovirus (Poliovirus) and Rhinovirus (Human rhinovirus, HRV), (2) Cardiovirus (Encephalomyocarditis virus, EMCV and Theiler's murine encephalomyelitis virus, TMEV) and Aphthovirus (Foot-and-mouth disease virus, FMDV), and (3) Hepatovirus (Hepatitis A virus, HAV) (reviewed in Belsham and Jackson, 2000; Kean *et al.*, 2001; Martínez-Salas and Fernández-Miragall, 2004).

Studies *in vitro* showed that the EMCV IRES-mediated initiation of translation is ATP dependent and requires eIF2, eIF3, eIF4A, and eIF4B as well as the central region of eIF4G to which eIF4A binds. eIF4E is not required, and therefore cleavage of eIF4G as well as the absence of eIF1 which is important for 40S ribosomal subunit scanning do not abolish EMCV IRES function. The same applies to the FMDV IRES-mediated translation from the first initiator AUG (reviewed in Pestova *et al.*, 2001). PTB, an auxiliary cellular 57-kDa protein with four RNA recognition motifs, strongly stimulates initiation of translation of all group 1 and 2 IRESs (Andreev *et al.*, 2007; Borovjagin *et al.*, 1994; Gosert *et al.*, 2000; Hunt and Jackson, 1999; Pilipenko *et al.*, 2000). ITAF45 is additionally required for FMDV IRES-mediated translation, and PCBP2 as well as unr for Poliovirus and Rhinovirus IRESs (Andreev *et al.*, 2007; Blyn *et al.*, 1997; Boussadia *et al.*, 2003; Pilipenko *et al.*, 2000). The La autoantigen stimulates PV IRES-mediated translation (Costa-Mattioli *et al.*, 2004).

The 5′ UTR of hepatovirus RNAs such as Hepatitis C virus (HCV; family Flaviviridae) are 342–385 nts long. Initiation at their IRES differs from initiation in picornavirus IRESs *in vitro*: binding of the 40S ribosomal subunit to the HCV IRES occurs directly, without requirement for the translation initiation factors eIF4F and eIF4B (reviewed in

Pestova *et al.*, 2001). Thus, the IRES functionally replaces eIF4F on the 40S ribosomal subunit (Siridechadilok *et al.*, 2005). Lack of eIF4F is compensated by conformational modifications in the 40S ribosomal subunit (Spahn *et al.*, 2001). Moreover, the activity of HCV-like IRESs is also affected by the coding sequence immediately downstream of the initiation codon. Not only flaviviruses but also RNA genomes of some picornaviruses such as Porcine Teschovirus carry HCV-like IRES elements within their 5′ UTR (Pisarev *et al.*, 2004).

An interesting variant of IRESs exists in viruses of the *Dicistroviridae* family such as Cricket paralysis virus (CrPV), viruses originally believed to be the insect counterpart of mammalian picornaviruses (reviewed in Doudna and Sarnow, 2007; Jan, 2006). The monopartite RNA genome of CrPV harbors a 5′ VPg and a 3′ poly(A) tail. It contains two nonoverlapping ORFs separated by an IGR (as opposed to picornaviruses that have only one ORF); the expression of the two ORFs is triggered by two distinct IRESs, one in the 5′ UTR and the other in the IGR (Jan *et al.*, 2003; Sasaki and Nakashima, 1999, 2000; Wilson *et al.*, 2000b). As shown using the long (580 nts) 5′ IRES contained in the Rhopalosiphum padi virus (RhPV) RNA, the 5′ UTR initiates translation of a nonstuctural polyprotein at the expected AUG codon. No specific boundaries of this IRES can be defined, suggesting that the IRES contains multiple domains capable of recruiting ribosomes for translation (Terenin *et al.*, 2005). The IGR of 175–533 nts separating the two ORFs contains the second IRES ($\sim$ 180 nts) that initiates synthesis of a structural polyprotein on a non-AUG codon and requires neither initiation factors nor Met-tRNAi^{Met} but a small domain (domain 3) downstream of the IGR IRES that docks into the 40S ribosomal P site mimicking the tRNA anticodon-loop structure during translation initiation (Costantino *et al.*, 2008; Jan and Sarnow, 2002; Wilson *et al.*, 2000a). In most cases, initiation from the second IRES begins at a GCU, GCA, or GCC triplet coding for alanine or at a CAA triplet coding for glutamine (reviewed in Pisarev *et al.*, 2005). In the model proposed for the initiation of the second IRES, domain 3 within the IGR occupies the ribosomal P site, the ribosomal A site remaining accessible for the Ala-tRNA or the Gln-tRNA, and translocation occurs on the ribosome without peptide bond formation (designated as the elongation-competent assembly of ribosome). As in the case of the HCV IRES, binding of the CrPV IGR to the 40S ribosomal subunit induces conformational changes on the ribosome (Costantino *et al.*, 2008; Pfingsten *et al.*, 2006; Spahn *et al.*, 2004).

The rates of cap- and IRES-dependent initiation pathways *in vitro* are different: using FMDV RNA as template it was shown that cap-dependent assembly of the 48S ribosomal complex occurs faster than IRES-mediated assembly (Andreev *et al.*, 2007). Moreover, some viruses have evolved sequences that prevent their IRESs from functioning. For example, the HCV IRES possesses a conserved stem–loop structure containing the initiation codon and this structure has been shown to decrease IRES efficiency (Honda *et al.*, 1996). One possible explanation is that for successful viral infection, IRESs should work at a very specific level of efficiency, which does not necessarily correspond to maximum efficiency.

As opposed to animal virus IRESs, plant virus IRES elements are shorter and less structured. Moreover, such elements are not confined to the 5′ UTRs on the genome of plant RNA viruses, and they are then at times referred to as TEs. Depending on the plant viral genome, the IRES is located (1) in the 5′ UTR such as in the picorna-like virus Potato virus Y (family Potyviridae; Levis and Astier-Manifacier, 1993), (2) within or between ORFs such as in a crucifer-infecting Tobacco mosaic virus (TMV, genus Tobamovirus; Jaag *et al.*, 2003; Skulachev *et al.*, 1999; Zvereva *et al.*, 2004) and in Potato leafroll virus (PLRV, family Luteoviridae; Jaag *et al.*, 2003), or (3) in the 3′ UTRs of viruses such as in BYDV (Guo *et al.*, 2001; Wang *et al.*, 1997). In the Hibiscus chlorotic ringspot virus (family Tombusviridae) genome, the activity of the 5′-located IRES is enhanced by the presence of a TE (also known as CITE) located in the 3′ UTR (Koh et al., 2002, 2003). The possible mechanisms of action of 3′ TEs has in recent years revealed the immense variety of strategies used in translation initiation by plant RNA viruses (reviewed in Miller and White, 2006).

E. Non-AUG initiation codons

In some cases in eukaryotes as also in prokaryotes, initiation of translation of cellular and viral mRNAs occurs on a non-AUG codon. Table I summarizes the situation for viruses that initiate some of their proteins on non-AUG codons.

In addition to containing the P ORF, the *Sendai virus* P mRNA harbors the C ORF in another reading frame that leads to the synthesis of a nested set of C-coterminal proteins (proteins C′, C, Y1, and Y2) known jointly as the C proteins (Fig. 2B). Except for C′ that is initiated upstream of the P protein on the mRNA, the other C proteins are entirely contained within the P ORF. C′ is initiated on an ACG codon in an optimum sequence context, and nts +5 and +6 also appear to be important for initiation at such non-AUG codons. The other C proteins (C, Y1, and Y2) are initiated on downstream-located AUG codons in suboptimal contexts and are presumably synthesized by leaky scanning or ribosome shunting (Curran and Kolakofsky, 1988; Gupta and Patwardhan, 1988; Kato *et al.*, 2004). Use of ACG as initiation codon has been described in the neurovirulent strains of TMEV (van Eyll and Michiels, 2002).

The initiation codon in *Human parainfluenza virus 1* (HPIV-1, family Paramyxoviridae) for the synthesis of C′ is a GUG codon. *In vivo*, GUG appears nearly as efficient as AUG in initiating C′ expression in the same context (Boeck *et al.*, 1992).

TABLE I Viruses shown or postulated to use non-AUG codons as initiators of protein synthesis

Family/genus	RNA[a]	Initiation codon	Protein	References
Plant viruses				
Caulimoviridae				
Tungrovirus				
RTBV	1	AUU	ORF1	Fütterer et al. (1996)
Furovirus				
SBWMV	2	CUG	28K[b]	Shirako (1998)
(Flexiviridae)				
(Foveavirus)				
PCMV	1	AUC	ORF1	James et al. (2007)
	1	AUA	ORF 5[b]	James et al. (2007)
Animal viruses				
Parvoviridae				
Dependovirus				
AAV-2	1	ACG	B	Becerra et al. (1985)
Retroviridae				
Lentivirus				
EIAV	1	CUG	Tat	Carroll and Derse (1993)
Gammaretrovirus				
MoMLV	1	CUG	Pr75^{gag}	Prats et al. (1989)
Deltaretrovirus				
HTLV-1	1	GUG	Rex	Corcelette et al. (2000)
		CUG	Tax	Corcelette et al. (2000)

Paramyxoviridae				
Respirovirus				
Sendai virus	1	ACG	C′	Boeck and Kolakofsky (1994), Curran and Kolakofsky (1988), and Gupta and Patwardhan (1988)
HPIV-1	1	GUG	C′	Boeck et al. (1992)
Picornaviridae				
Cardiovirus				
TMEV	1	AUG/ACG[c]	L*	van Eyll and Michiels (2002)
Flaviviridae				
Flavivirus				
HCV	1	GUG/GCG[c]	F	Baril and Brakier-Gingras (2005)
Dicistroviridae				
Cripavirus				
CrPV	1	GCU[d]	ORF 2	Wilson et al. (2000a)
PSIV	1	CAA[e]	ORF 2	Sasaki and Nakashima (2000) and Yamamoto et al. (2007)
RhPV	1	GCA[d]	ORF 2	Domier et al. (2000)

For each virus, the RNA segment whose protein is initiated at a non-AUG codon is indicated as also the initiation codon used, and the designation of the resulting protein. The brackets surrounding Flexiviridae and Foveavirus indicate that PCMV is presumed to belong to this family and genus.

[a] RTBV contains a double-stranded DNA; AAV-2 contains a single-stranded DNA.
[b] CP ORF.
[c] Depending on the variant.
[d] Ala as initiator.
[e] Gln as initiator.

The Moloney murine leukemia virus (MoMLV, family Retroviridae) genomic RNA codes for two in-phase precursor proteins Pr65^{gag} and Pr75^{gag}. It uses an upstream CUG as translation initiation codon for the synthesis of Pr75^{gag} that migrates to the cell surface and is involved in virus spread (Prats *et al.*, 1989).

Among plant viruses, a non-AUG initiation codon exists in the poly-cistronic mRNA of the pararetrovirus Rice tungro bacilliform virus (RTBV; family Caulimoviridae). ORF I of the mRNA (harboring ORFs I–III) is accessed by reinitiation after translation of a short ORF (sORF). Following a long 5′ leader sequence harboring several sORFs that are bypassed by ribosome shunting, synthesis is initiated on an AUU codon at ORF I (Fütterer and Hohn, 1996). Only 10% of the ribosomes initiate at ORF I; the remaining 90% reach ORFs II and III and initiate on an AUG codon (reviewed in Ryabova *et al.*, 2006). A similar situation occurs in other pararetroviruses. RNA 2 of the bipartite (+) sense single-stranded RNA genome of Soil-borne wheat mosaic virus (SBWMV, genus Furovirus) codes for two proteins; the shorter (19K) CP is produced via conventional AUG initiation, whereas the N-terminally extended 28K protein is initiated at a CUG codon upstream of the CP ORF (Shirako, 1998). Under certain conditions, AUU codons located in the 5′ UTR of the TMV RNA can serve as initiation codons (Schmitz *et al.*, 1996).

In the cases presented earlier, the non-AUG codons allow initiation with a methionine residue. There is however an interesting situation of methionine-independent translation initiation (reviewed in Pisarev *et al.*, 2005; Touriol *et al.*, 2003). This is the case of the IRES-dependent initiation of translation of members of the Dicistroviridae family whose structural protein encoded by ORF 2 lacks an AUG initiation codon and translation initiation occurs at a CAA (coding for Gln) or GCU or GCA (coding for Ala) codon, depending on the virus (Table I).

F. Multiple reading frames

Whereas eukaryotic cell mRNAs are usually monocistronic, the mRNAs of eukaryotic viruses frequently contain several ORFs, the AUG posi-tioned close to the 5′ end of the RNA generally constituting the initiation codon. To reach downstream initiation codons that correspond to internal ORFs on polycistronic RNAs lacking an IRES, viruses resort to either leaky scanning, reinitiation, or shunting.

1. Leaky scanning

A mechanism commonly used by viruses to express polycistronic RNAs is leaky scanning (reviewed in Ryabova *et al.*, 2006), in which when the initiation codon lies within less than 10 nts from the cap structure, or when it is embedded in a poor context for initiation, some of the scanning ribosomes bypass this first initiation codon and start translation on a downstream-located initiation codon whose context is more appropriate for initiation (Fig. 2B). Leaky scanning also occurs when initiation is at a non-AUG codon in an optimal context followed by an AUG codon. Two possible situations can arise: in-frame initiation or overlapping ORFs.

a. In-frame initiation   This occurs when an ORF harbors more than one potential in-frame initiation codon; it is codon context-dependent. The outcome of in-frame initiation is the production of two proteins that are identical over the total length of the shorter protein. Table II lists the cases of in-frame initiation reported. In FMDV and Plum pox potyvirus (PPV, family Potyviridae), in-frame initiation is cap independent (Andreev *et al.*, 2007; Simon-Buela *et al.*, 1997).

b. Overlapping ORFs   This strategy is extremely common among viruses and is generally also codon context-dependent. The result of this strategy is the synthesis of two different proteins. A situation common to plant

TABLE II   In-frame initiation

Family/genus	Genome segment	Protein	References
Comoviridae			
Comovirus			
CPMV	RNA M	Movement protein	Verver *et al.* (1991)
Hordevirus			
BSMV	RNA β	Movement protein	Petty and Jackson (1990)
Furovirus			
SBWMV	RNA 2	Coat protein	Shirako (1998)
Potyviridae			
Potyvirus			
PPV	RNA	Polyprotein	Simon-Buela *et al.* (1997)
Bornaviridae			
Bornavirus			
BDV	RNA P	24- and 16-kDa phosphoproteins	Kobayashi *et al.* (2000)
Picornaviridae			
Aphthovirus			
FMDV	RNA	Polyprotein	Andreev *et al.* (2007)

For each virus the genome segment that undergoes in-frame initiation is indicated.

viruses belonging to several genera such as the carlaviruses and potex-viruses (family Flexiviridae), and the viruses of the genera Furovirus and Hordeivirus is the presence within their (+) sense single-stranded RNA genome of a group of three ORFs known as the triple gene block whose expression leads to three proteins involved in movement of the virus within the plant. Synthesis of these proteins requires the production of two sgRNAs. The 5′-proximal ORF is translated from a functionally monocistronic sgRNA, whereas the two subsequent ORFs are translated from the second sgRNA. Expression of the third ORF, which overlaps the second ORF, occurs by leaky scanning and is codon context-dependent (Verchot *et al.*, 1998; Zhou and Jackson, 1996).

Peanut clump virus (PCV, genus Pecluvirus) contains a bipartite (+) sense single-stranded strand RNA genome. In RNA2, the first of two ORFs that codes for the virus CP terminates with a UGA codon that overlaps the AUG codon initiating the second ORF: AUGA. About one-third of the ribosomes fail to initiate translation of the CP and scan the template initiating translation of the second ORF, more than 100 residues downstream of the first ORF (Herzog *et al.*, 1995). RTBV contains a closed-circular double-stranded DNA genome that is transcribed yielding two mRNAs. The longer polycistronic mRNA (known as pregenomic RNA) encodes three ORFs (I, II, and III) that are linked by AUGA, the termina-tion codon of the upstream ORF overlapping the initiation codon of the downstream ORF (Fütterer *et al.*, 1997). ORF I is initiated at an AUU codon, preceded by a long 5′ UTR with several sORFs that are bypassed by ribosome shunting. On the other hand, ORFs II and III initiate at a conventional AUG codon. However, the AUG initiating ORF II is in a poor context, and the majority of the ribosomes bypass this AUG to reach the downstream more favorable AUG of ORF III. Leaky scanning there-fore accounts for initiation of translation of ORFs II and III.

Turnip yellow mosaic virus (family Tymoviridae) is a monopartite (+) sense single-stranded RNA virus that bears a cap structure, and harbors a tRNA-like structure (TLS) at its 3′end that can be valylated *in vitro* and *in vivo*. Its first two 5′-proximal and largely overlapping ORFs code for the movement protein (ORF1), and the replicase polyprotein (ORF2) in a different reading frame. It has been reported that the valylated viral RNA serves as bait for ribosomes directing them to initiate synthesis of ORF2, and donating its valine residue for the N-terminus of the polypro-tein in a cap- and initiator-independent manner (Barends *et al.*, 2003); interaction between the 3′ TLS and the initiation codon of ORF2 would lead to circularization of the RNA. However, recent studies suggest that initiation of translation of the polyprotein is cap and context dependent, the TLS having only a positive effect on translation of ORF2 without being indispensable (Matsuda and Dreher, 2007). This mechanism allows dicis-tronic expression from initiation codons that are closely spaced.

2. Reinitiation

Another possibility for initiation at an internal start codon in a polycistronic mRNA is reinitiation of translation of downstream ORFs following expression of a 5′-proximal ORF (of 30 codons or less; reviewed in Ryabova *et al.*, 2006). Reinitiation requires that the 40S ribosomal subunit remain on the mRNA after terminating synthesis of the 5′-proximal ORF. Efficiency of reinitiation decreases with increasing length of the IGR between the 5′-proximal and the next ORF.

Among eukaryotic viruses, polycistronic mRNAs have been the most thoroughly examined in viruses of the family Caulimoviridae, in particular in the double-strand DNA virus Cauliflower mosaic virus (CaMV). The large 35S mRNA of CaMV and related viruses contains up to seven ORFs (Fig. 5), and for some of them recurrent translation depends on reinitiation activated by the transactivator (TAV). The TAV protein is encoded by ORF VI contained in the pregenomic (or polycistronic) 35S mRNA; it is expressed by the 19S sgRNA in which it is the only ORF (Pooggin *et al.*, 2001). In dicistronic constructs harboring the CaMV ORF VII followed by ORF I (or by ORFs II, III, IV, V, or an artificial ORF) fused to the chloramphenicol acetyltransferase (CAT) gene, very low levels of CAT activity were obtained in plant protoplasts; however, when the

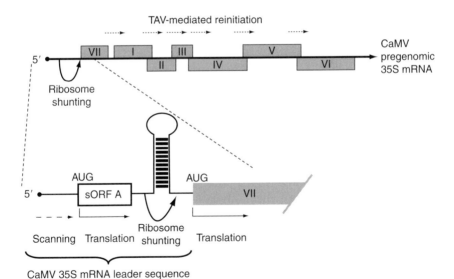

FIGURE 5 Schematic representation of the organization of the CaMV pregenomic 35S mRNA and strategies of translation initiation. I–VII are ORFs. TAV, transactivator. Arrows show migration of ribosomes by reinitiation (dotted), scanning (dashed), and shunting (curved). Translation is represented by a bent arrow. Other indications are as in legend of Fig. 1.

product of ORF VI was included, considerably higher levels of CAT activity were observed (Bonneville *et al.*, 1989; Fütterer and Hohn, 1991). The second ORF of the dicistronic construct is synthesized by reinitiation and not by an IRES, since a stem structure positioned at various sites upstream of this ORF hinders its translation (Fütterer and Hohn, 1991). TAV-stimulated initiation of the second ORF does not depend on the distance separating the two ORFs, since the distance can be abolished as in a quadruplet AUGA, or the ORFs can be separated by as many as 700 nts, and even limited overlap between the ORFs is possible. TAV directly binds to the eIF3g subunit of eIF3 and associates with the L18 and L24 proteins of the 60S ribosomal subunit (Leh *et al.*, 2000; Park *et al.*, 2001). These interactions result in TAV–eIF3 complex association with the trans-locating ribosome during translation, favoring reinitiation of downstream ORFs. On the other hand, eIF4B can compete with TAV for binding to eIF3g, since the binding sites of these two proteins on eIF3g overlap. Overexpression of eIF4B inhibits TAV-mediated reinitiation of a second ORF, probably by inhibiting TAV–eIF3g-40S complex formation (Park *et al.*, 2004).

The members of the *Calicivirus* family contain a (+) sense single-stranded RNA carrying a VPg. The sgRNAs of these viruses that also contain a VPg represent widely studied examples of reinitiation by mammalian ribosomes after translation of a long ORF. The Rabbit hemorrhagic disease virus genomic RNA codes for a large polyprotein ORF1 that is subsequently processed producing the viral nonstructural proteins and the 3′ terminally located major CP VP60, as well as a small 3′ terminally located ORF2 in another reading frame. The 3′-terminal part of ORF1 overlaps the 5′ region of ORF2. Expression of ORF2 yields the minor CP VP10 and is produced from a sgRNA that also contains the region of ORF1 expressing VP60. Thus, the sgRNA codes for the major VP60 encoded by the 3′-terminal part of ORF1, and for the minor VP10 produced by ORF2. The two ORFs overlap by AUGUCGA such that the termination codon (UGA) of ORF1 lies downstream of the initiation codon (AUG) of ORF2. Synthesis of VP10 occurs from the genomic as well as from the sgRNA and involves an unusual translation termination/reinitiation process. Indeed, synthesis of VP10 depends strictly on the presence of the termination codon ending ORF1 preceded by a sequence element of about 80 nts (Meyers, 2003). The sequence element contains two motifs that are essential for expression of ORF2, one of which is conserved among caliciviruses and is complementary to a sequence in the 18S ribosomal RNA. In FCV, sgORF1 and sgORF2 overlap by 4 nts (AUGA) and translation in a reticulocyte lysate of the FCV sgRNAs showed that ORF1/ORF2 termination/reinitiation does not require the eIF4F complex and that the 3′-terminal RNA sequence of ORF1 binds to the 40S ribosomal subunit and to IF3 (Luttermann and Meyers, 2007;

Meyers, 2007; Pöyry *et al.*, 2007). Thus, the termination/reinitiation process requires sequence elements that could prevent dissociation of postterminating ribosomes via RNA–RNA, RNA–protein, and/or protein–protein interactions.

3. Shunting

A ribosome shunting mechanism has been proposed to explain how initiation of translation occurs in viral polycistronic mRNAs that have a long leader sequence with generally several sORFs, a long low-energy hairpin structure and a probable packaging signal within the 5′ UTR (reviewed in Ryabova *et al.*, 2006). This is the case of CaMV (Fig. 5). The ribosomes having entered at the level of the cap structure on the 35S mRNA would scan a few nts, then skip from a "take-off site" over part of the leader sequence containing a structural element and sORFs, to reach a "landing site," and finally scan to the downstream ORF. It has been suggested that formation of a leader hairpin between the two sites would bring these sites in close proximity, favoring shunting. It is generally assumed that shunting is more easily achieved if the upstream ORF is short, such that the initiation factors that allowed initiation of translation of the sORF may have at least partly remained on the ribosome during translation (reviewed in Jackson, 2005). In addition to the size of the sORF, the time required for scanning seems also to be important (Pöyry *et al.*, 2004), the eIF4F initiation complex remaining on the ribosome for a few seconds without interruption of sORF translation. The leader sequence of the CaMV 35S mRNA is replete with sORFs. Of these, the 5′-proximal sORF, sORF A, is indispensable for ribosome shunting and infectivity; its aa sequence is generally not important but it must be translated and should be between 2 and 10 codons long for efficient shunting. Another important *cis*-acting element for shunting includes the distance between the termination codon of sORF A and the base of the leader hairpin (reviewed in Ryabova *et al.*, 2006). Finally, it has been reported that TAV promotes expression of ORF VII (Pooggin *et al.*, 2001).

Shunting may explain translation of polycistronic mRNAs in other viruses, generally by examining the effect on translation of inserting a strong hairpin structure near the 5′ end or in the middle region of the leader sequence, or by inserting AUG codons within the leader, as done for CaMV. Shunting occurs in the case of the 200 nt-long leader, the tripartite leader, in the Adenovirus late mRNAs from the major late promoter. This highly conserved leader contains a 25–44 nt-long unstructured 5′ region, followed by highly structured hairpins devoid of sORFs. Shunting has been reported to be enhanced by complementarity between the tripartite leader and the 3′ hairpin of the 18S ribosomal RNA (Xi *et al.*, 2004; Yueh and Schneider, 2000).

In the polycistronic P/C mRNA of Sendai virus (Fig. 2B), proteins P and C are presumably initiated by leaky scanning, whereas proteins Y1 and Y2 most possibly arise by shunting. This was suggested because changing the ACG codon of C' to AUG dramatically reduced the synthesis of the P and C proteins, but had virtually no effect on the synthesis of Y1 and Y2 (Latorre *et al.*, 1998). Yet to date, no specific sites have been detected in the mRNA to account for shunting.

G. Modification of cell factors involved in initiation

Shutoff of host protein synthesis is the process in which cell protein synthesis is inhibited during viral infection due to the use by the virus of the host metabolism (reviewed in Gale *et al.*, 2000; Randall and Goodbourn, 2008). Host shutoff reflects the competition between viral and host mRNAs for the translation machinery, and results in selective translation of viral mRNAs over endogenous host mRNAs. Early translational switch is accompanied by disaggregation of polysomes containing capped cellular mRNAs, followed by reformation of polysomes containing exclusively viral mRNAs (reviewed in Lloyd, 2006).

It is at first sight rather surprising that in plants, no infection by a plant virus has so far been conclusively demonstrated to hinder host translation *in planta* so as to favor synthesis of viral proteins. Host translational shutoff by plant viruses has been reported only in *in vitro* translation studies of the potyviruses TuMV and TEV (Cotton *et al.*, 2006; Khan *et al.*, 2008; Miyoshi *et al.*, 2006). The authors reported different causes for the inhibition of cellular mRNA translation. On one hand inhibition would be the result of competition between cellular-capped mRNAs and VPg for eIFiso4E, the binding affinity of VPg for eIFiso4E being stronger that of the capped mRNA (Khan *et al.*, 2008; Miyoshi *et al.*, 2006). On the other hand, inhibition of cell mRNA translation by TuMV would not be mediated by the interaction of VPg-Pro (precursor of VPg) with eIFiso4E but by VP-Pro-induced degradation of RNAs (Cotton *et al.*, 2006).

It has been established for several plants that variation in eIF4E and eIFiso4E is involved in natural recessive resistance against potyviruses (reviewed in Kang *et al.*, 2005; Robaglia and Caranta, 2006). Resistance and complementation assays provide evidence for coevolution between pepper eIF4E and potyviral VPg (Charron *et al.*, 2008). Some recessive plant virus resistance genes code for eIF4E with the aa substitution Gly107Arg, and this substitution was shown to abolish the ability of eIF4E to bind TEV VPg and the cap, providing resistance against TEV infection (Yeam *et al.*, 2007). Recently, a functional map of lettuce eIF4E was obtained, and the results using mutated eIF4E suggest that the function of eIF4E in the potyvirus cycle might be distinct from its physiological function of binding the cap structure at the 5′ ends of mRNAs to initiate translation; thus eIF4E may be required for virus RNA replication or other processes of the virus cycle (German-Retana *et al.*, 2008).

1. Phosphorylation of eIF2α

The function of eIF2 in protein synthesis is the formation of the TC and its delivery to the 40S ribosomal subunit. eIF2 is a complex composed of the three subunits α, β, and γ (Fig. 6). Phosphorylation of eIF2α inhibits the exchange of GDP for GTP catalyzed by the exchange factor eIF2B, and leads to the sequestration of eIF2B in a complex with eIF2 resulting in general inhibition of protein synthesis (Sudhakar *et al.*, 2000; reviewed in Hinnebusch, 2005). The amount of eIF2B in cells is limiting as compared to eIF2. Thus, even small changes in the phosphorylation status of eIF2α have a drastic effect on translation due to eIF2B sequestration

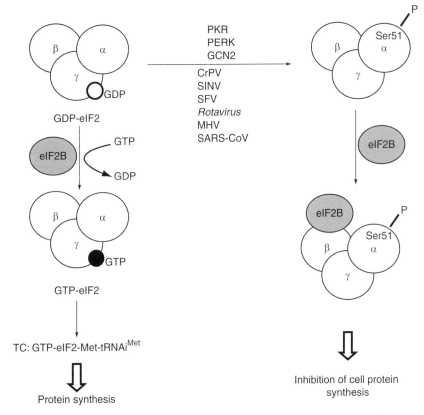

FIGURE 6   Phosphorylation of eIF2α in virus-infected cells. TC, ternary complex; P, phosphorylation. α, β, and γ are the subunits of eIF2.

(Balachandran and Barber, 2004; Krishnamoorthy *et al.*, 2001; Sudhakar *et al.*, 2000; Yang and Hinnebusch, 1996). For several mRNAs the eIF2 complex is replaced by a single polypeptide designated eIF2A that directs codon-dependent and GTP-independent Met-tRNAi^{Met} binding to the 40S ribosomal subunit and may act by favoring expression of specific proteins (Adams *et al.*, 1975; Merrick and Anderson, 1975; Zoll *et al.*, 2002).

Four cellular eIF2α kinases are known to phosphorylate the eIF2α subunit at residue Ser51. Three of the kinases—the protein kinase RNA (PKR), the PKR-like endoplasmic reticulum kinase (PERK), and the general control nonderepressible-2 (GCN2) kinase—play a prominent role in virus-infected cells (Fig. 6). PKR binds to and is activated by double-strand RNAs that are generated during replication and transcription of viral genomes. Accumulation of unfolded proteins in the endoplasmic reticulum during viral infection induces a signaling cascade from the cytoplasmic kinase domain of PERK, leading to induction of eIF2α phosphorylation. Finally, GCN2 kinase is reported to be activated upon Sindbis virus (SINV, family Togaviridae) infection (Berlanga *et al.*, 2006).

Many viruses evolved diverse strategies to prevent PKR or PERK activation in infected cells; these strategies have been discussed in detail in recent reviews (Dever *et al.*, 2007; Garcia *et al.*, 2007; Mohr, 2006; Mohr *et al.*, 2007). However, there are several examples in which viruses use eIF2α phosphorylation to switch off cell translation and direct the cell machinery to synthesize their own proteins (Fig. 6). A classical illustration of how eIF2 modification fosters translation of viral mRNAs is initiation of translation on the CrPV IRES. The IRES contained in the IGR promotes initiation of protein synthesis without the assistance of any initiation factors, including eIF2 (reviewed in Doudna and Sarnow, 2007; Pisarev *et al.*, 2005). Moreover, CrPV stimulates eIF2α phosphorylation; this inactivates host mRNA translation by decreasing the amount of preinitiation 43S ribosomal complexes formed and facilitates translation initiation on the CrPV IRES. Indeed, lowering the amounts of TC and 43S ribosomal complexes increases the efficiency of initiation on the CrPV IRES (Pestova *et al.*, 2004; Thompson *et al.*, 2001). HCV encodes proteins known to inactivate PKR (or PKR + PERK) function(s) (Garcia *et al.*, 2007). However, HCV IRES-driven translation initiation can also be maintained in the presence of activated PKR and reduced TCs (Robert *et al.*, 2006). A new pathway of eIF2- and eIF5-independent initiation of translation on the HCV IRES has been proposed recently in which assembly of the 80S complex requires only two initiation factors, eIF5B and eIF3 (Terenin *et al.*, 2008).

Infection by viruses of the genus Alphavirus (family Togaviridae) such as SINV or Semliki forest virus (SFV) activates PKR, which results in almost complete phosphorylation of eIF2α at late times postinfection. Translation of the viral sg 26S mRNA takes place efficiently during this time, whereas translation of genomic mRNA is impaired by eIF2α phosphorylation (Molina *et al.*, 2007; Ventoso *et al.*, 2006). It was shown that a hairpin loop structure within the 26S mRNA-coding region, located downstream of the AUG initiation codon, promotes eIF2-independent translation with the help of eIF2A (Ventoso *et al.*, 2006). However, the fact that translation of the 26S mRNA must be coupled to transcription to be efficient in infected cells suggests that additional viral or cellular factors are involved in translation initiation on the 26S mRNA (Sanz *et al.*, 2007).

Early in the infection process rotaviruses take over the host translation machinery, and this is achieved via interaction of the viral NSP3 with eIF4G and phosphorylation of eIF2α (Figs. 4C and 6; Montero *et al.*, 2008; Piron *et al.*, 1998). These two mechanisms may explain the severe shutoff of cell protein synthesis observed during rotavirus infection, although it is not clear how capped viral mRNAs are efficiently translated in such eIF2α-sequestered conditions.

Murine hepatitis virus (MHV) as well as Severe acute respiratory syndrome coronavirus (SARS-CoV), both of the family Coronaviridae, induce host translational shutoff. This is achieved via different mechanisms: degradation of cell mRNAs including mRNAs encoding translation-related factors (Leong *et al.*, 2005; Raaben *et al.*, 2007), increase in eIF2α phosphorylation presumably via PERK, and formation of stress granules and processing bodies that are thus sites of mRNA stalling and degradation, respectively (Chan *et al.*, 2006; Raaben *et al.*, 2007; Versteeg *et al.*, 2006). Expression of the SARS-CoV NSP1 is involved in degradation of several host mRNAs and in host translation shutoff (Kamitani *et al.*, 2006). Surprisingly, despite eIF2α phosphorylation the SARS-CoV proteins are still efficiently synthesized even though coronaviral mRNAs are structurally equivalent to host mRNAs (Hilton *et al.*, 1986; Siddell *et al.*, 1981).

It is interesting to observe that despite considerable work performed in recent years, phosphorylation of eIF2α still represents one of the most intriguing problems in translational control during viral infection, since it is still not clear why the phosphorylation of eIF2 affects cellular protein synthesis without impairing translation initiation of many viral RNAs.

2. Modification of eIF4E and 4E-BP

eIF4E is believed to be the least abundant of all initiation factors and, therefore, to be a perfect target for regulation of protein synthesis. It interacts with the cap structure of mRNAs, with the scaffold protein eIF4G and with repressor proteins known as eIF4E-binding proteins (4E-BPs). eIF4E undergoes regulated phosphorylation on Ser209 mediated by the eIF4G-associated MAPK signal-integrating kinases, Mnk1 and Mnk2 (Fig. 7) (Pyronnet *et al.*, 1999; Raught and Gingras, 2007). Uninfected cells growing exponentially typically possess roughly equal amounts of phosphorylated and nonphosphorylated forms of eIF4E (Feigenblum and Schneider, 1993) and the ratio

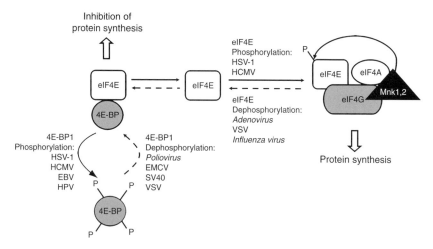

FIGURE 7 Phosphorylation and dephosphorylation of eIF4E and 4E-BP1. eIF4E+eIF4G+eIF4A form eIF4F. Other indications are as in legends of Figs. 1 and 6.

shifts toward the phosphorylated form of eIF4E following treatment of the cells with growth factors, hormones, and mitogens (Flynn and Proud, 1995; Joshi *et al.*, 1995; Makkinje *et al.*, 1995). However, the functional role of eIF4E phosphorylation remains elusive. Indeed, there is no direct link between eIF4E phosphorylation and the enhanced translation observed as a result of these stimuli, since recent studies showed that phosphorylation of eIF4E decreases the affinity of eIF4E for capped mRNA. Thus, the working hypothesis is that the nonphosphorylated form of eIF4E within the eIF4F complex (eIF4E, eIF4G, and eIF4A) binds to the cap structure on the mRNA, and that eIF4E phosphorylation accompanies initiation complex transition to elongation (reviewed in Scheper and Proud, 2002). In addition, phosphorylation could dissociate eIF4E from the cap and enable the eIF4F complex to move along the 5′ UTR and unwind the secondary structure.

4E-BP constitutes a family of translation repressors that prevent eIF4F assembly and act as negative growth regulators (Raught and Gingras, 2007). 4E-BPs are phosphoproteins, 4E-BP1 being the best studied of the three 4E-BPs known in mammals. It undergoes phosphorylation at multiple sites leading to its dissociation from eIF4E, leaving eIF4E free to bind eIF4G and to form the eIF4F complex (Fig. 7) (Lin *et al.*, 1994; Pause *et al.*, 1994). The mechanism proposed is that eIF4E possesses an eIF4G-binding site which overlaps with 4E-BP motifs; thus, 4E-BP and eIF4G binding to eIF4E would be mutually exclusive (Haghighat *et al.*, 1995; Marcotrigiano *et al.*, 1999).

a. Dephosphorylation of eIF4E and of 4E-BP1   Adenovirus (family Adenoviridae), Vesicular stomatitis virus (VSV; family Rhabdoviridae), and Influenza virus infections lead to accumulation of nonphosphorylated eIF4E and subsequent inhibition of host protein synthesis. Adenovirus mediates the quantitative dephosphorylation of eIF4E (up to 95% of the total eIF4E) leading to suppression of cellular protein synthesis (Fig. 7) (Feigenblum and Schneider, 1993). The Adenovirus late protein designated 100K is synthesized at high levels at the onset of the late phase of infection (Bablanian and Russell, 1974; Oosterom-Dragon and Ginsberg, 1980). It interacts with the C-terminus of eIF4G (Cuesta *et al.*, 2000) and with the tripartite leader sequence of viral late mRNAs (Xi *et al.*, 2004). Binding of the 100K protein to eIF4G evicts Mnk1 from the eIF4F complex, thus impairing eIF4E phosphorylation in the initiation complex and inhibiting translation of host mRNAs (Cuesta *et al.*, 2000). On the other hand, adenoviral late mRNAs are translated efficiently via ribosome shunting (Xi et al., 2004, 2005). VSV infection causes dephosphorylation of eIF4E and 4E-BP1 thus hampering host protein synthesis (Fig. 7). The resulting changes in eIF4F do not inhibit translation of viral mRNAs, although the detailed mechanism of how VSV mRNAs that are capped and possess poly(A) tails overcome the obstacle created has not been elucidated (Connor and Lyles, 2002). Influenza virus infection results in partial (up to 70%) dephosphorylation of eIF4E and concomitant loss of eIF4F activity (Fig. 7). Thus, Influenza virus mRNAs that are capped via cap-snatching and polyadenylated (Herz *et al.*, 1981; Krug *et al.*, 1979; Luo *et al.*, 1991) are translated efficiently under conditions of partial inactivation of eIF4F (Feigenblum and Schneider, 1993) when host protein synthesis is blocked (Katze and Krug, 1990). Several studies have shown that the NS1 viral protein selectively promotes translation of viral mRNAs by increasing their rate of initiation (de la Luna *et al.*, 1995; Enami *et al.*, 1994; Katze *et al.*, 1986; Park and Katze, 1995) and interacts with PABP and eIF4GI (one of the two isoforms of eIF4G in animals) in viral mRNA translation initiation complexes (Aragón *et al.*, 2000; Burgui *et al.*, 2003). Moreover, a recent report has provided evidence that the Influenza virus RdRp substitutes for eIF4E in viral mRNA translation and binds to the translation preinitiation complex (Burgui *et al.*, 2007). One can speculate that the combination of dephosphorylation of eIF4E, hyperphosphorylation of eIF4G, and binding of RdRp to the preinitiation complex and of NS1 to eIF4GI creates an eIF4F factor more specific for Influenza virus mRNA translation.

4E-BP1 is dephosphorylated following infection with Poliovirus or EMCV (Fig. 7). This is a well-established example of viral switch from cap-dependent to IRES-mediated initiation of translation in picornavirus-infected cells (Gingras *et al.*, 1996; Svitkin *et al.*, 2005). Simian virus 40 (SV40; family Polyomaviridae) is a recent example of a virus that causes significant decrease in phosphorylation of 4E-BP1 late in lytic infection. This process is specifically mediated by the SV40 small t antigen. As in the case of Poliovirus and EMCV, dephosphorylation of 4E-BP1 and its subsequent binding to eIF4E displaces eIF4E from the eIF4F complex. This mechanism functions as a switch in translation initiation mechanisms favoring IRES-mediated translation (Yu *et al.*, 2005). Indeed, recent studies have shown that the SV40 late 19S mRNA possesses an IRES (Yu and Alwine, 2006).

b. Phosphorylation of eIF4E and 4E-BP1   Members of the Herpesviridae family of the Alphaherpesvirinae subfamily, such as Herpes Simplex Virus 1 (HSV-1), and of the Betaherpesvirinae subfamily, such as Human cytomegalovirus (HCMV), can stimulate the assembly of eIF4F complexes in primary human cells; this is partly achieved by phosphory-lation of eIF4E and 4E-BP1 early in the productive viral growth cycle (Fig. 7) (Kudchodkar *et al.*, 2006; Walsh and Mohr, 2004; Walsh *et al.*, 2005). At the same time HSV-1 infection dramatically impairs host protein synthesis (Elgadi *et al.*, 1999; Everly *et al.*, 2002; Sciabica *et al.*, 2003) whereas with HCMV the effect on host protein synthesis is weak (Stinski, 1977). Interestingly, the ratio of eIF4F over 4E-BP1 increases in cells infected with either HSV-1 or HCMV, promoting assembly of eIF4F complexes. For HSV-1 this is achieved exclusively through proteasome-mediated degrada-tion of 4E-BP1 (Walsh and Mohr, 2004), whereas for HCMV, replication induces an increase in the overall abundance of the eIF4F components eIF4E and eIF4G, and also of PABP relative to the translational repressor 4E-BP1 (Walsh *et al.*, 2005). However, liberation of eIF4E from 4E-BP1 in the case of HSV-1 is insufficient to accelerate eIF4E incorporation into the eIF4F com-plex. A recent study showed that the HSV-1 ICP6 gene product binds to eIF4G promoting association of eIF4E with the N-terminus of eIF4G and facilitating eIF4E phosphorylation. This suggests a chaperone role for ICP6 in eIF4F assembly (Walsh and Mohr, 2006).

4E-BP1 is hyperphosphorylated (Fig. 7) following infection by *Epstein–Barr Virus* (EBV; family Herpesviridae, subfamily Gammaherpesvirinae) (Moody *et al.*, 2005) or Human papillomavirus (HPV; family Papilloma-viridae) (Moody *et al.*, 2005; Munger *et al.*, 2004; Oh *et al.*, 2006).

3. Modification of eIF4G
a. Cleavage of eIF4G   The large modular protein eIF4G serves as a dock-ing site for initiator factors and other proteins involved in initiation of RNA translation. Due to the central role of eIF4G in translation initiation, many viruses belonging to the families Picornaviridae, Retroviridae, and Caliciviridae have evolved mechanisms to modify the function of eIF4G so as to prevent cell protein synthesis. These viruses induce cleavage of eIF4G, separating the N-terminal eIF4E-binding domain from the C-terminal eIF4A- and eIF3-binding domains (Fig. 8). As a consequence, the capacity of eIF4G to connect capped mRNAs to the 40S ribosome is abolished by the virus, inducing host translation shutoff.

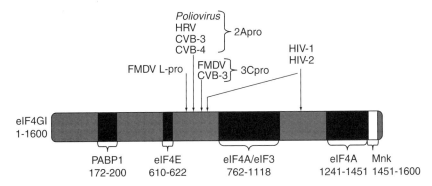

FIGURE 8   eIF4GI protein-binding sites and cleavage sites by viral proteases. Black or white rectangles represent regions of eIF4G interacting with other proteins. Numbers correspond to aa. Arrows correspond to sites of cleavage.

Host shutoff during infection by picornaviruses such as Poliovirus, HRV and human Coxsackie virus B (CVB)-3 and CVB-4 results in part from cleavage of eIF4GI by the viral 2A protease (2Apro) at aa 681/682 (Baxter *et al.*, 2006; Lamphear *et al.*, 1993; Sommergruber *et al.*, 1994; Sousa *et al.*, 2006). The Poliovirus and HRV 2Apro also cleave eIF4GII (at aa 699/670) but more slowly than cleavage of eIF4GI (Gradi *et al.*, 1998; Svitkin *et al.*, 1999). FMDV has evolved an alternate papain-like protease, L-pro in place of 2Apro to cleave both isoforms of eIF4G (Gradi *et al.*, 2004); it cleaves eIF4GI 7 aa upstream of 2Apro (Fig. 8), and eIF4GII 1 aa downstream of the 2Apro cleavage site (reviewed in Lloyd, 2006). Poliovirus infection also activates two cell proteases that cleave eIF4GI close to the 2Apro cleavage site (Zamora *et al.*, 2002). The 3Cpro of FMDV and the 2Apro and 3Cpro of CVB-3 also cleave eIF4GI (Fig. 8) (Chau *et al.*, 2007; Strong and Belsham, 2004). Degradation of eIF4GI has been observed in CD4+ cells infected with Human immunodeficiency virus 1 (HIV-1; family Retroviridae) (Ventoso *et al.*, 2001). The HIV-1 protease efficiently cleaves eIF4GI at multiple sites, but not eIF4GII (Ohlmann *et al.*, 2002). Proteases of HIV-2 and of members of the family Retroviridae (Human T-lymphotropic virus 1 (HTLV-1), Simian immunodeficiency virus, and Mouse mammary tumor virus (MMTV)) also cleave eIF4GI (Alvarez *et al.*, 2003; reviewed in Lloyd, 2006). Finally, infection of cells with FCV leads to cleavage of eIF4GI and eIF4GII and host translation shutoff (Willcocks *et al.*, 2004); the identity of the protease responsible for cleavage of eIF4G is unknown, but it could be a cellular protease activated by the infection.

b. Phosphorylation of eIF4G   eIF4G is 10-fold more phosphorylated in Influenza virus-infected than in noninfected cells and phosphorylated eIF4G still interacts with eIF4A and eIF4E. Cleavage of eIF4G by the

Poliovirus 2Apro inhibits translation of the Influenza virus mRNAs (Feigenblum and Schneider, 1993; Garfinkel and Katze, 1992). Phosphorylation of eIF4G in HCMV-infected cells is one of the mechanisms that enhances eIF4F activity during the viral replication cycle (Kudchodkar *et al.*, 2004; Walsh *et al.*, 2005). eIF4G phosphorylation is induced throughout infection with SV40 (Yu *et al.*, 2005).

4. Cleavage of PABP

Certain viruses cleave the C-terminal domain of PABP thereby destroying its interactions with eIF4B, eRF3, or Paip1 (Fig. 4A). PABP is targeted for cleavage by the 2Apro and 3Cpro of Poliovirus and CVB-3 (Joachims *et al.*, 1999; Kerekatte *et al.*, 1999; Kuyumcu-Martinez et al., 2002, 2004b), by L-pro of FMDV (Rodríguez Pulido *et al.*, 2007), and by 3Cpro of HAV (Zhang *et al.*, 2007). PABP is proteolytically processed by the calicivirus 3C-like protease (Kuyumcu-Martinez *et al.*, 2004a), and HIV-1 and HIV-2 proteases are also able to cleave PABP in the absence of other viral proteins (Alvarez *et al.*, 2006). The contribution of PABP cleavage versus eIF4G cleavage in shutoff of host or viral protein synthesis has not been compared directly. Poliovirus cleavage of PABP appears to be promoted by the interaction of PABP with translation initiation factors, ribosomes or poly(A)-containing RNAs (Kuyumcu-Martinez *et al.*, 2002; Rivera and Lloyd, 2008). Processing of PABP could either occur through one of the components that provides shutoff of host translation or could favor the switch from translation to replication of viral genomes as for example PABP cleavage by 3Cpro in Poliovirus- and HAV-infected cells (Bonderoff *et al.*, 2008; Zhang *et al.*, 2007).

5. Substitution of PABP

Severe inhibition of host mRNA translation due to competition between the viral protein NSP3 and PABP for eIF4G was shown in cells infected with rotaviruses. The viral NSP3 protein binds to the conserved motif UGACC located at the 3' end of the viral mRNA, and circularizes the mRNA via interaction with eIF4G (Fig. 4C). Since NSP3 has a higher affinity for eIF4G than does PABP, it replaces PABP and disrupts host mRNA circularization (Michel *et al.*, 2000; Vende *et al.*, 2000). NSP3–eIF4G interaction results in reduced efficiency of host mRNA translation. NSP3-mediated circularization has been reported to enhance Rotavirus mRNA translation (Vende *et al.*, 2000) and to be dispensable for translation of the viral mRNAs (Montero *et al.*, 2006). X-ray structure and biophysical studies have shown that NSP3 forms an asymmetric homodimer around the conserved motif at the 3' end of Rotavirus mRNAs (Deo *et al.*, 2002).

6. Cleavage of PBCP2

During the mid-to-late phase of *Poliovirus* infection, PCBP2 is cleaved by the viral proteases 3C and 3CD (Fig. 4B); the cleaved protein is no longer able to bind to the IRES and initiate translation, but it binds to the 5′-terminal cloverleaf structure or simultaneously to the cloverleaf structure and to the adjacent C-rich spacer circularizing the viral genome for replication. Hence, the formation of two different closed loop structures could favor the switch from translation to replication of the *Poliovirus* genome (Gamarnik and Andino, 1998; Herold and Andino, 2001; Perera *et al.*, 2007; Toyoda *et al.*, 2007).

IV. ELONGATION OF TRANSLATION

A. Frameshift

This is the mechanism whereby during the course of peptide chain elongation, certain ribosomes shift from the original ORF (0 frame) on the mRNA by one nt, either in the 5′ direction (-1 frame) or in the 3′ direction ($+1$ frame), and continue protein synthesis in the new frame. This results in the synthesis of two proteins, the "stopped" and "transframe" proteins; they are identical from the N-terminus to the frameshift site but differ thereafter, and the stopped protein is always the more abundant of the two proteins (reviewed in Farabaugh, 2000). The occurrence of -1 frameshift is more frequent and has been more extensively studied than $+1$ frameshift. -1 Frameshift is common among $(+)$ strand RNA viruses; it has been found in most retroviruses, in coronaviruses, L-A viruses of yeast and in several plant viruses belonging to diverse groups (Table III). Frameshift is observed during translation of RNA genomes exhibiting overlapping gene arrangements. It usually allows the expression of the viral replicase, the transframe protein in most cases harboring the polymerase or the reverse transcriptase. It has recently been reported (Chung *et al.*, 2008) that TuMV, in addition to synthesizing a large polyprotein that undergoes cleavage, also harbors a frameshift protein embedded in the P3 region of the polyprotein: frameshift leads to the expression of a protein designated P3-PIPO. PIPO is essential for infectivity of the virus, although its precise role has not been established.

Three RNA signals are important in -1 frameshifting, a slippery heptanucleotide sequence where frameshift occurs, a downstream hairpin that in many instances can additionally form a pseudoknot, and a spacer element between the slippery sequence and the hairpin structure; the length of the spacer varies between 4 and 9 nts, depending on the viral genome. The viral sequences appear to be optimized for a suitable level of stopped and transframe proteins required for viral replication rather than for maximum frameshift (Kim *et al.*, 2001).

TABLE III   Viruses of eukaryotes shown or postulated to regulate elongation of translation by frameshifting

Family/genus	RNA	Type of FS	Proteins	References
Plant viruses[a]				
Carlavirus				
PVM	1	−1	CP/12K	Gramstat *et al.* (1994)
Sobemovirus				
BWYV	1	−1	66K/67K	Veidt *et al.* (1988, 1992)
CoMV	1	−1	64K/56K	Mäkinen *et al.* (1995)
Closteroviridae				
Closterovirus				
BYV	1	+1	295K/48K	Agranovsky *et al.* (1994)
CCSV	1	+1	ORF1a/b	ten Dam *et al.* (1990)
CTV	1	+1	349K/57K	Karasev *et al.* (1995)
Crinivirus				
LIYV	1	+1	217K/55K	Klaassen *et al.* (1995)
Luteoviridae				
Enamovirus				
PEMV	1	−1	84K/67K	Demler and de Zoeten (1991)
	2	−1	33K/65K	Demler *et al.* (1993)
Luteovirus				
BYDV-PAV	1	−1	39K/60K	Di *et al.* (1993)
Polerovirus				
PLRV	1	−1	70K/67K	Prüfer *et al.* (1992)
Tombusviridae				
Dianthovirus				
CRSV	1	−1	27K/54K	Kujawa *et al.* (1993) and Ryabov *et al.* (1994)
RCNMV	1	−1	27K/57K	Kim and Lommel (1994) and Xiong *et al.* (1993)
SCNMV	1	−1	27K/57K	Ge *et al.* (1993)
Animal viruses				
Pseudoviridae[b]				
Pseudovirus				
SceTy1V	1	+1	Gag/Pol	Belcourt and Farabaugh (1990) and Clare *et al.* (1988)

(continued)

TABLE III   (*continued*)

Family/genus	RNA	Type of FS	Proteins	References
Metaviridae[b]				
Metavirus				
SceTy3V	1	+1	Gag/Pol	Hansen *et al.* (1992)
Errantivirus				
DmeGypV	1	−1	Gag/Pol	Bucheton (1995)
Retroviridae[b]				
Alpharetrovirus				
ASLV	1	−1	Pro/Pol	Arad *et al.* (1995)
Betaretrovirus				
MMTV	1	−1	Gag/Pro/ Pol	Jacks *et al.* (1987)
Deltaretrovirus				
HTLV-1	1	−1	Gag/Pro/ Pol	Nam *et al.* (1993)
Lentivirus				
HIV-1	1	−1	Gag/Pro	Parkin *et al.* (1992)
Totiviridae[c]				
Totivirus				
ScV-L-A	1	−1	Gag/Pol	Dinman *et al.* (1991)
Leishmaniavirus				
LRV1-1	1	+1	CP/RdRp	Stuart *et al.* (1992)
Astroviridae[a]				
Astrovirus				
HAstV	1	−1	ORF1a/ ORF1b	Lewis and Matsui (1996)
Coronaviridae[a]				
Coronavirus				
IBV	1	−1	Pol1a/ Pol1b	Brierley *et al.* (1987, 1989)
SARS-CoV				Plant and Dinman (2006)
Torovirus				
EqTV	1	−1	ORF1a/ ORF1b	Lai and Cavanagh (1997)
Arteriviridae[a]				
Arterivirus				
EAV	1	−1	ORF1a/ ORF1b	den Boon *et al.* (1991) and Napthine *et al.* (2003)

(*continued*)

TABLE III   (*continued*)

Family/genus	RNA	Type of FS	Proteins	References
Unassigned virus[a]				
APV	1	− 1	ORF1/ ORF2	van der Wilk *et al.* (1997)
Measles virus	1	− 1	P/R	Liston and Briedis (1995)
BLV				Rice *et al.* (1985)

For each virus the genome segments that undergo frameshifting (FS), the type of FS, and the "stopped" and "transframe" proteins involved are indicated.
[a] (+) Sense single-stranded RNA viruses.
[b] Reverse-transcribing RNA viruses.
[c] dsRNA viruses.

B. Modification of elongation factors

eEF-1 is composed of eEF1A (formerly called eEF-1α) the transporter of aa-tRNAs to the A site on the ribosomes during elongation in conjunction with GTP hydrolysis, and a trimeric complex known as eEF1B (formerly called eEF-1βγδ) responsible for the regeneration of GTP from GDP on eEF-1A (Slobin and Moller, 1978). eEF-2 promotes translocation of the aa- or peptidyl-tRNA from the A to the P site on the ribosome in a GTP-dependent reaction.

Given the fact that strong evidence for deviations from the norm during elongation of protein synthesis does not seem to exist, it is not surprising that the cases of modification of elongation factors caused by viral infection appear to be virtually nonexistent. Indeed, such modifications would most likely equally affect cellular and viral protein synthesis. Nevertheless, a case of elongation factor modification has been documented during infection by viruses of the *Herpesviridae* family.

The mammalian eEF-1δ subunit of eEF1B is phosphorylated *in vitro* in the same position by the cell kinase cdc2, and hyperphosphorylated by a viral kinase conserved in all the subfamilies of the *Herpesviridae* family, such as the HSV-1 UL13 kinase, the EBV BGLF4 kinase, and the HCMV UL97 kinase (Kato *et al.*, 2001; Kawaguchi *et al.*, 1999, 2003). How phosphorylation of eEF-1δ by the viral kinases affects translation elongation remains obscure.

V. TERMINATION OF TRANSLATION

Termination of translation occurs when the ribosome encounters one of the three termination codons that defines the 3′ boundary of the ORF on the mRNA: UAG, UGA, or UAA. It involves termination codon recognition at the ribosomal A site, peptidyl-tRNA hydrolysis, and release of ribosomes from the mRNA. The participation of two proteins, the eukaryotic release factors eRF1 and eRF3, in termination codon recognition has been demonstrated (Drugeon *et al.*, 1997; Janzen *et al.*, 2002; Karamysheva *et al.*, 2003). The three termination codons are decoded by eRF1 that catalyzes ester bond hydrolysis in peptidyl-tRNA at the ribosomal peptidyl-transferase center. eRF1 functions cooperatively with the GTPase eRF3 whose activity is ribosome and eRF1 dependent (Kononenko *et al.*, 2008; Pisareva *et al.*, 2006). Final events leading to complete disassembly of the posttermination 80S ribosome require eIF1, eIF1A, and eIF3 (Pisarev *et al.*, 2007). Efficiency of termination appears to be determined by competition between eRF binding to the ribosome and alternative translational events that allow ribosomes to continue decoding. The processes that can circumvent termination codons include: ribosomal frameshift, readthrough or suppression of termination by natural cellular tRNAs, and binding of release factors.

A. Readthrough

In readthrough, a cellular aa-tRNA, called a natural suppressor, decodes the termination codon and translation continues in the same frame up to the next in-frame termination codon. Readthrough is commonly encountered in plant single-stranded RNA viruses and in some animal viruses. Table IV presents the families, genera, and viruses whose genomes have been shown or postulated to resort to readthrough. Readthrough usually allows the synthesis of the RdRp, the reverse transcriptase or of a CP-fusion protein, depending on the virus. The CP-fusion protein is present in the virus particles and is needed for encapsidation and/or for vector transmission.

Readthrough of termination codons requires the positioning of a suppressor aa-tRNA in the ribosomal A site where it competes with eRF1 for the termination codon. Two proteins are produced in the presence of a suppressor aa-tRNA that recognizes the termination codon at the 3′ end of an ORF: the expected "stopped" protein that terminates at the termination codon of the ORF, and the longer "readthrough" protein that extends to the next in-frame termination codon. The two proteins are identical over the total length of the stopped protein. Synthesis of the stopped protein is always more abundant than that of the readthrough protein.

TABLE IV   Viruses shown or postulated to regulate termination of translation by readthrough

Family/genus	RNA	Termination codon	Proteins	References
Plant viruses[a]				
Benyvirus				
BNYVV	2	UAG	CP/75K	Niesbach-Klosgen *et al.* (1990) and Schmitt *et al.* (1992)
Furovirus				
SBWMV	1	UGA	150K/209K	Shirako and Wilson (1993)
	2	UGA	CP/84K	Yamamiya and Shirako (2000)
Peclovirus				
PCV	1	UGA	103K/191K	Herzog *et al.* (1994)
Pomovirus				
BVQ	1	UAA	149K/207K	Koenig *et al.* (1998)
	2	UAG	CP/54K	Koenig *et al.* (1998)
BSBV	1	UAA	145K/204K	Koenig and Loss (1997)
	2	UAG	CP/104K	Koenig *et al.* (1997)
Tobamovirus				
TMV	1	UAG	126K/183K	Ishikawa *et al.* (1986), Pelham (1978), and Skuzeski *et al.* (1991)
Tobravirus				
PEBV	1	UGA	141K/201K	MacFarlane *et al.* (1989)
TRV	1	UGA	134K/194K	Hamilton *et al.* (1987)
Tombusviridae				
Avenavirus				
OCSV	1	UAG	23K/84K	Boonham *et al.* (1995)
Carmovirus				
CarMV	1	UAG,UAG	27K/86K/ 98K	Guilley *et al.* (1985)
CCFV	1	UAG	28K/87K	Skotnicki *et al.* (1993)
MNSV	1	UAG	29K/89K	Riviere and Rochon (1990)
		UAG	7K/14K	Riviere and Rochon (1990)

(continued)

TABLE IV   (*continued*)

Family/genus	RNA	Termination codon	Proteins	References
TCV	1	UAG	28K/88K	White *et al.* (1995)
Machlomovirus				
MCMV	1	UAG	50K/111K	Nutter *et al.* (1989)
		UGA	9K/33K	Nutter *et al.* (1989)
Necrovirus				
TNV	1	UAG	23K/82K	Meulewaeter *et al.* (1990)
Tombusvirus				
AMCV	1	UAG	33K/92K	Tavazza *et al.* (1994)
CNV	1	UAG	33K/92K	Rochon and Tremaine (1989)
CyRSV	1	UAG	33K/92K	Grieco *et al.* (1989)
TBSV	1	UAG	33K/92K	Hearne *et al.* (1990)
Luteoviridae				
Enamovirus				
PEMV	1	UGA	CP/55K	Demler and de Zoeten (1991)
SbDV	1	UAG	CP/80K	Rathjen *et al.* (1994)
Luteovirus				
BYDV-PAV	1	UAG	CP/72K	Dinesh-Kumar *et al.* (1992), Filichkin *et al.* (1994), Miller *et al.* (1988), and Wang *et al.* (1995)
Polerovirus				
BWYV	1	UAG	CP/74K	Brault *et al.*, 1995 and Veidt *et al.*, 1988, 1992
PLRV	1	UAG	CP/80K	Bahner *et al.* (1990) and Rohde *et al.* (1994)
Animal viruses				
Retroviridae[b]				
Gammaretrovirus				
MLV	1	UAG	Gag/Pol	Etzerodt *et al.* (1984) and Herr (1984)
Epsilonretrovirus				
WDSV	1	UAG	Gag/Pro	Holzschu *et al.* (1995)

(*continued*)

TABLE IV   *(continued)*

Family/genus	RNA	Termination codon	Proteins	References
Togaviridae[a]				
Alphavirus				
SINV	1	UGA	P123/nsP4	Strauss and Strauss (1994)

For each virus, the RNA segment whose protein undergoes readthrough, the nature of the suppressible termination codon, and the designation of the stopped (indicated as CP or by its size if not the CP) and readthrough proteins (indicated by the total size of the resulting protein) are indicated. Other members of the Alphavirus genus (O'nyong-nyong virus and SFV) have CGA (Arg); one SINV strain has UGU (Cys); in all cases, the importance of a C residue 3′ of UGA, CGA, or UGU has been emphasized.
[a] (+) Sense single-stranded RNA viruses.
[b] Reverse-transcribing RNA viruses.

A well-known example of readthrough occurs in the TMV (+) single-stranded RNA genome. The 5′-proximal ORF codes for the 126K protein that contains a putative methyltransferase and a helicase domain. Readthrough of its UAG termination codon leads to the synthesis of the 183K readthrough product that harbors the highly conserved GDD (Gly-Asp-Asp) motif, responsible for replicase activity (reviewed in Beier and Grimm, 2001; Maia *et al.*, 1996).

Many members of the genus Alphavirus harbor a suppressible UGA codon separating the regions coding for the NSP3 and NSP4 proteins. The NSP4 protein shares homologous aa sequences with the RdRp of *Poliovirus* and plant RNA viruses.

In addition to the termination codon, other *cis* elements on the mRNA are required for efficient readthrough. These elements are either the sequence surrounding the termination codon preferentially on the 3′ side and/or a hairpin or pseudoknot structure also located downstream of the suppressible termination codon. In the case of TMV RNA, the nature of the two codons following the suppressible UAG codon affects the level of readthrough (Valle *et al.*, 1992). The requirements in BYDV are very different: two elements are mandatory for readthrough of the UAG codon *in vitro* and *in vivo*: a proximal and a distal element located, respectively, 6–15 nts and about 700 nts downstream of the suppressible UAG codon (Brown *et al.*, 1996). The distal element is conserved among luteoviruses and in Pea enation mosaic enamovirus (PEMV, family Luteoviridae), suggesting that it might also participate in readthrough in these viruses.

Readthrough was clearly demonstrated in mouse cells infected with Murine leukemia virus (MLV; family Retroviridae). Here, most ribosomes terminate synthesis at the UAG codon to produce the Gag protein, but when termination is suppressed, a glutamine residue from Gln-tRNA[Gln] is incorporated at the level of the UAG codon and elongation continues to form the Gag–Pol product. This latter protein is then cleaved to yield Gag, a protease (whose corresponding gene segment harbors the suppressed UAG codon) and the reverse transcriptase (Yoshinaka *et al.*, 1985). In retroviruses, suppression of termination is controlled by structures within the RNA itself: it requires a few specific nts immediately downstream of the termination codon, followed by a spacer region of a few nts and a hairpin that in some cases forms a pseudoknot. In MLV, suppression of the *gag* UAG codon depends on specific downstream sequences and on a pseudoknot structure (reviewed in Gale *et al.*, 2000).

B. Suppressor tRNAs

Misreading of termination codons is achieved by a variety of naturally occurring suppressor tRNAs that normally recognize a cognate codon, but at times recognize one of the termination codons by "improper" base pairing (reviewed in Beier and Grimm, 2001).

1. Suppressors of UAG/UAA codons

The first natural UAG suppressor tRNA identified was the cytoplasmic tRNA^{Tyr} bearing a GΨA anticodon purified from tobacco leaves and *Drosophila melanogaster* (Beier *et al.*, 1984; Bienz and Kubli, 1981). Pseudouridine (Ψ) can form a classical base pair with adenosine. The Ψ modification at the second anticodon position is necessary to read the UAG codon; it enhances the unconventional G:G interaction at the first anticodon position. Mutating the suppressible TMV UAG codon to UAA leads to virion formation in plants, implying that a tRNA recognizing the UAA codon is present in the host. It was shown that the UAA codon, if placed in the TMV context, was also recognized *in vitro* by the suppressor tobacco tRNA^{Tyr}. A second UAG/UAA suppressor is the cytoplasmic tRNA^{Gln} with CUG or UmUG (Um is 2′-O-methyluridine) anticodons. tRNA^{Gln} is present in almost all prokaryotes and eukaryotes. Interaction of the two tRNA^{Gln} isoacceptors with UAG or UAA requires an unconventional G:U base pair at the third anticodon position. Probably an unmodified A in the tRNA immediately 3′ of the anticodon facilitates noncanonical base pairing. Other UAG suppressors are the cytoplasmic tRNA^{Leu} with a CAA or a CAG anticodon. Here, recognition of the UAG codon requires an unusual A:A pair in the second position of both the CAA and the CAG anticodons and also a G:U pair in the third position of the CAG anticodon.

2. Suppressors of UGA codons

Two UGA suppressors, a chloroplast and a cytoplasmic tRNA^{Trp} with the anticodon CmCA (Cm is 2′-O-methylcytidine) were isolated from tobacco plants and shown to suppress the Tobacco rattle virus (TRV; genus, Tobravirus) RNA1 UGA codon. Several reports indicate that a tRNA^{Trp}

with UGA suppressor activity is also present in vertebrates (Cordell *et al.*, 1980; Geller and Rich, 1980). Recognition of the UGA codon by tRNA^{Trp} requires an unusual Cm:A pair in the first position of the CmCA anticodon. A tRNA^{Cys} with a GCA anticodon was isolated from tobacco plants and shown to suppress the UGA in TRV RNA1. Misreading of UGA by tRNA^{Cys} involves a G:A pair at the first GCA anticodon position. The two tRNA^{Arg} with an U*CG (U* is 5-methoxy-carbonylmethyluridine) or ICG anticodon stimulate UGA readthrough in the context of TRV RNA1. Interaction of tRNA^{Arg} with the UGA codon requires a G:U base pair at the third U*CG anticodon position.

C. Binding of release factors

The reverse transcriptase of MLV interacts with eRF1. This interaction displaces eRF3 from the release factor complex and increases synthesis of the readthrough protein. This function of the reverse transcriptase is required for appropriate levels of the readthrough and stopped proteins (Orlova *et al.*, 2003; reviewed in Goff, 2004).

Interaction between the nascent peptidyl-tRNA during translation of the 22-codon upstream ORF2 (uORF2) and eRF1 of HCMV inhibits expression of the downstream UL4 gene. The peptide product of uORF2 inhibits its own translation termination by forming a stable peptidyl-prolyl-tRNA-ribosome complex that prevents peptide release and stalls the elongating ribosome at the uORF2 termination codon (Janzen *et al.*, 2002).

VI. CONCLUSIONS

The study of the regulation of gene expression has known various phases over the decades, ever since some of its major players, such as messenger RNAs and ribosomes had been identified. It first led to examining the initiation, elongation, and termination steps of protein biosynthesis using bacterial extracts and artificial RNAs or bacteriophage RNA genomes as mRNAs, and defining the proteins involved in each step. Thereafter, the availability of cell mRNAs greatly facilitated the study of protein biosynthesis in extracts of eukaryotic cells. This revealed the vast number of protein factors involved in particular at the initiation step of protein synthesis, and the mechanism of action of these factors. In recent years, the sequencing of an ever increasing number of viral RNA genomes shown to function as mRNAs has brought a wealth of new information regarding the fundamental role played by the modulation of the structure of mRNAs in regulating gene expression. It has, for instance, led to numerous studies that consider circularization of mRNAs an important step in promoting protein synthesis. In addition, it has also highlighted the variety of strategies developed by viruses to perturb host protein synthesis so as to favor synthesis of viral proteins. Such evasion of host protein synthesis is now leading to a variety of fascinating studies showing that this involves a complex yet balanced interplay between the host cell translation machinery, the viral mRNA, and the viral proteins resulting from expression of the viral genome. Further experiments will undoubtedly unveil other new venues in this intriguing and multifaceted aspect of cell development.

ACKNOWLEDGMENTS

We are very grateful to Hans Gross for suggesting that we write a review on the present topic, and for his encouragements. We are most thankful to Ivan N. Shatsky and Nahum Sonenberg for critical reading of the review and for their valuable and constructive suggestions. We are most grateful to Katherine M. Kean for her support during the writing of this article.

This work was supported by DARRI (Direction des Applications de la Recherche et des Relations Industrielles de l'Institut Pasteur) to A. V. K. and by grants from SIDACTION and the ANRS (Association Nationale de Recherche contre le SIDA) to B. C. R.

A. V. K. thanks her father, Vasiliy B. Poltaraus, for supporting and educating her. The authors thank the members of their laboratories for their help and constant support.

REFERENCES

Adams, S. L., Safer, B., Anderson, W. F., and Merrick, W. C. (1975). Eukaryotic initiation complex formation. Evidence for two distinct pathways. *J. Biol. Chem.* **250:**9083–9089.

Agranovsky, A. A., Koonin, E. V., Boyko, V. P., Maiss, E., Frotschl, R., Lunina, N. A., and Atabekov, J. G. (1994). Beet yellows closterovirus: Complete genome structure and identification of a leader papain-like thiol protease. *Virology* **198:**311–324.

Alvarez, E., Castello, A., Menendez-Arias, L., and Carrasco, L. (2006). HIV protease cleaves poly(A)-binding protein. *Biochem. J.* **396:**219–226.

Alvarez, E., Menendez-Arias, L., and Carrasco, L. (2003). The eukaryotic translation initiation factor 4GI is cleaved by different retroviral proteases. *J. Virol.* **77:**12392–12400.

Andreev, D. E., Fernandez-Miragall, O., Ramajo, J., Dmitriev, S. E., Terenin, I. M., Martínez-Salas, E., and Shatsky, I. N. (2007). Differential factor requirement to assemble translation initiation complexes at the alternative start codons of foot-and-mouth disease virus RNA. *RNA* **13:**1366–1374.

Arad, G., Bar-Meir, R., and Kotler, M. (1995). Ribosomal frameshifting at the Gag-Pol junction in avian leukemia sarcoma virus forms a novel cleavage site. *FEBS Lett.* **364:**1–4.

Aragón, T., de la Luna, S., Novoa, I., Carrasco, L., Ortín, J., and Nieto, A. (2000). Eukaryotic translation initiation factor 4GI is a cellular target for NS1 protein, a translational activator of influenza virus. *Mol. Cell. Biol.* **20:**6259–6268.

Bablanian, R., and Russell, W. C. (1974). Adenovirus polypeptide synthesis in the presence of non-replicating poliovirus. *J. Gen. Virol.* **24:**261–279.

Bahner, I., Lamb, J., Mayo, M. A., and Hay, R. T. (1990). Expression of the genome of potato leafroll virus: Readthrough of the coat protein termination codon *in vivo*. *J. Gen. Virol.* **71**:2251–2256.

Balachandran, S., and Barber, G. N. (2004). Defective translational control facilitates vesicular stomatitis virus oncolysis. *Cancer Cell* **5**:51–65.

Balvay, L., Lopez Lastra, M., Sargueil, B., Darlix, J. L., and Ohlmann, T. (2007). Translational control of retroviruses. *Nat. Rev. Microbiol.* **5**:128–140.

Barends, S., Bink, H. H., van den Worm, S. H., Pleij, C. W., and Kraal, B. (2003). Entrapping ribosomes for viral translation: tRNA mimicry as a molecular Trojan horse. *Cell* **112**:123–129.

Baril, M., and Brakier-Gingras, L. (2005). Translation of the F protein of hepatitis C virus is initiated at a non-AUG codon in a $+1$ reading frame relative to the polyprotein. *Nucleic Acids Res.* **33**:1474–1486.

Barr, J. N. (2007). Bunyavirus mRNA synthesis is coupled to translation to prevent premature transcription termination. *RNA* **13**:731–736.

Baxter, N. J., Roetzer, A., Liebig, H. D., Sedelnikova, S. E., Hounslow, A. M., Skern, T., and Waltho, J. P. (2006). Structure and dynamics of coxsackievirus B4 2A proteinase, an enyzme involved in the etiology of heart disease. *J. Virol.* **80**:1451–1462.

Becerra, S. P., Rose, J. A., Hardy, M., Baroudy, B. M., and Anderson, C. W. (1985). Direct mapping of adeno-associated virus capsid proteins B and C: A possible ACG initiation codon. *Proc. Natl. Acad. Sci. USA* **82**:7919–7923.

Bedard, K. M., Daijogo, S., and Semler, B. L. (2007). A nucleo-cytoplasmic SR protein functions in viral IRES-mediated translation initiation. *EMBO J.* **26**:459–467.

Beier, H., Barciszewska, M., Krupp, G., Mitnacht, R., and Gross, H. J. (1984). UAG readthrough during TMV RNA translation: Isolation and sequence of two tRNAs with suppressor activity from tobacco plants. *EMBO J.* **3**:351–356.

Beier, H., and Grimm, M. (2001). Misreading of termination codons in eukaryotes by natural nonsense suppressor tRNAs. *Nucleic Acids Res.* **29**:4767–4782.

Belcourt, M. F., and Farabaugh, P. J. (1990). Ribosomal frameshifting in the yeast retrotransposon Ty: tRNAs induce slippage on a 7 nucleotide minimal site. *Cell* **62**:339–352.

Belsham, G. J., and Jackson, R. J. (2000). Translational initiation on picornavirus RNA. *In* "Translational Control of Gene Expression" (N. Sonenberg, J. W. B. Hershey, and M. B. Mathews, eds.), pp. 869–900. CSHL Press, Cold Spring Harbor, NY.

Berlanga, J. J., Ventoso, I., Harding, H. P., Deng, J., Ron, D., Sonenberg, N., Carrasco, L., and de Haro, C. (2006). Antiviral effect of the mammalian translation initiation factor 2alpha kinase GCN2 against RNA viruses. *EMBO J.* **25**:1730–1740.

Bernardi, F., and Haenni, A. L. (1998). Viruses: Exquisite models for cell strategies. *Biochimie* **80**:1035–1041.

Bienz, M., and Kubli, E. (1981). Wild-type tRNATyrG reads the TMV RNA stop codon, but Q base-modified tRNATyrQ does not. *Nature* **294**:188–190.

Blyn, L. B., Towner, J. S., Semler, B. L., and Ehrenfeld, E. (1997). Requirement of poly (rC) binding protein 2 for translation of poliovirus RNA. *J. Virol.* **71**:6243–6246.

Boeck, R., Curran, J., Matsuoka, Y., Compans, R., and Kolakofsky, D. (1992). The parainfluenza virus type 1 P/C gene uses a very efficient GUG codon to start its C′ protein. *J. Virol.* **66**:1765–1768.

Boeck, R., and Kolakofsky, D. (1994). Positions $+5$ and $+6$ can be major determinants of the efficiency of non-AUG initiation codons for protein synthesis. *EMBO J.* **13**:3608–3617.

Bol, J. F. (2005). Replication of alfamo- and ilarviruses: Role of the coat protein. *Annu. Rev. Phytopathol.* **43**:39–62.

Bonderoff, J. M., Larey, J. L., and Lloyd, R. E. (2008). Cleavage of Poly(A)-binding protein by poliovirus 3C proteinase inhibits viral internal ribosome entry site-mediated translation. *J. Virol.* **82:**9389–9399.

Bonneville, J. M., Sanfacon, H., Fütterer, J., and Hohn, T. (1989). Posttranscriptional trans-activation in cauliflower mosaic virus. *Cell* **59:**1135–1143.

Boonham, N., Henry, C. M., and Wood, K. R. (1995). The nucleotide sequence and proposed genome organization of oat chlorotic stunt virus, a new soil-borne virus of cereals. *J. Gen. Virol.* **76:**2025–2034.

Borovjagin, A., Pestova, T., and Shatsky, I. (1994). Pyrimidine tract binding protein strongly stimulates *in vitro* encephalomyocarditis virus RNA translation at the level of preinitiation complex formation. *FEBS Lett.* **351:**299–302.

Bouloy, M., Plotch, S. J., and Krug, R. M. (1978). Globin mRNAs are primers for the transcription of influenza viral RNA *in vitro*. *Proc. Natl. Acad. Sci. USA* **75:**4886–4890.

Boussadia, O., Niepmann, M., Creancier, L., Prats, A. C., Dautry, F., and Jacquemin-Sablon, H. (2003). Unr is required *in vivo* for efficient initiation of translation from the internal ribosome entry sites of both rhinovirus and poliovirus. *J. Virol.* **77:**3353–3359.

Brault, V., van den Heuvel, J. F., Verbeek, M., Ziegler-Graff, V., Reutenauer, A., Herrbach, E., Garaud, J. C., Guilley, H., Richards, K., and Jonard, G. (1995). Aphid transmission of beet western yellows luteovirus requires the minor capsid read-through protein P74. *EMBO J.* **14:**650–659.

Brierley, I., Boursnell, M. E., Binns, M. M., Bilimoria, B., Blok, V. C., Brown, T. D., and Inglis, S. C. (1987). An efficient ribosomal frame-shifting signal in the polymerase-encoding region of the coronavirus IBV. *EMBO J.* **6:**3779–3785.

Brierley, I., Digard, P., and Inglis, S. C. (1989). Characterization of an efficient coronavirus ribosomal frameshifting signal: Requirement for an RNA pseudoknot. *Cell* **57:**537–547.

Brierley, I., and Dos Ramos, F. J. (2006). Programmed ribosomal frameshifting in HIV-1 and the SARS-CoV. *Virus Res.* **119:**29–42.

Brown, C. M., Dinesh-Kumar, S. P., and Miller, W. A. (1996). Local and distant sequences are required for efficient readthrough of the barley yellow dwarf virus PAV coat protein gene stop codon. *J. Virol.* **70:**5884–5892.

Bucheton, A. (1995). The relationship between the flamenco gene and gypsy in Drosophila: How to tame a retrovirus. *Trends Genet.* **11:**349–353.

Burgui, I., Aragón, T., Ortín, J., and Nieto, A. (2003). PABP1 and eIF4GI associate with influenza virus NS1 protein in viral mRNA translation initiation complexes. *J. Gen. Virol.* **84:**3263–3274.

Burgui, I., Yanguez, E., Sonenberg, N., and Nieto, A. (2007). Influenza virus mRNA translation revisited: Is the eIF4E cap-binding factor required for viral mRNA translation? *J. Virol.* **81:**12427–12438.

Bushell, M., and Sarnow, P. (2002). Hijacking the translation apparatus by RNA viruses. *J. Cell Biol.* **158:**395–399.

Bushell, M., Wood, W., Carpenter, G., Pain, V. M., Morley, S. J., and Clemens, M. J. (2001). Disruption of the interaction of mammalian protein synthesis eukaryotic initiation factor 4B with the poly(A)-binding protein by caspase- and viral protease-mediated cleavages. *J. Biol. Chem.* **276:**23922–23928.

Carroll, R., and Derse, D. (1993). Translation of equine infectious anemia virus bicistronic tat-rev mRNA requires leaky ribosome scanning of the tat CTG initiation codon. *J. Virol.* **67:**1433–1440.

Casey, J. L. (2002). RNA editing in hepatitis delta virus genotype III requires a branched double-hairpin RNA structure. *J. Virol.* **76:**7385–7397.

Cattaneo, R., Kaelin, K., Baczko, K., and Billeter, M. A. (1989). Measles virus editing provides an additional cysteine-rich protein. *Cell* **56:**759–764.

Chan, C. P., Siu, K. L., Chin, K. T., Yuen, K. Y., Zheng, B., and Jin, D. Y. (2006). Modulation of the unfolded protein response by the severe acute respiratory syndrome coronavirus spike protein. *J. Virol.* **80:**9279–9287.

Charron, C., Nicolai, M., Gallois, J. L., Robaglia, C., Moury, B., Palloix, A., and Caranta, C. (2008). Natural variation and functional analyses provide evidence for co-evolution between plant eIF4E and potyviral VPg. *Plant J.* **54:**56–68.

Chau, D. H., Yuan, J., Zhang, H., Cheung, P., Lim, T., Liu, Z., Sall, A., and Yang, D. (2007). Coxsackievirus B3 proteases 2A and 3C induce apoptotic cell death through mitochondrial injury and cleavage of eIF4GI but not DAP5/p97/NAT1. *Apoptosis* **12:**513–524.

Cheng, Q., Jayan, G. C., and Casey, J. L. (2003). Differential inhibition of RNA editing in hepatitis delta virus genotype III by the short and long forms of hepatitis delta antigen. *J. Virol.* **77:**7786–7795.

Chisholm, G. E., and Henner, D. J. (1988). Multiple early transcripts and splicing of the Autographa californica nuclear polyhedrosis virus IE-1 gene. *J. Virol.* **62:**3193–3200.

Christensen, A. K., Kahn, L. E., and Bourne, C. M. (1987). Circular polysomes predominate on the rough endoplasmic reticulum of somatotropes and mammotropes in the rat anterior pituitary. *Am. J. Anat.* **178:**1–10.

Chung, B. Y., Miller, W. A., Atkins, J. F., and Firth, A. E. (2008). An overlapping essential gene in the Potyviridae. *Proc. Natl. Acad. Sci. USA* **105:**5897–5902.

Clare, J. J., Belcourt, M., and Farabaugh, P. J. (1988). Efficient translational frame-shifting occurs within a conserved sequence of the overlap between the two genes of a yeast Ty1 transposon. *Proc. Natl. Acad. Sci. USA* **85:**6816–6820.

Connor, J. H., and Lyles, D. S. (2002). Vesicular stomatitis virus infection alters the eIF4F translation initiation complex and causes dephosphorylation of the eIF4E binding protein 4E-BP1. *J. Virol.* **76:**10177–10187.

Corcelette, S., Masse, T., and Madjar, J. J. (2000). Initiation of translation by non-AUG codons in human T-cell lymphotropic virus type I mRNA encoding both Rex and Tax regulatory proteins. *Nucleic Acids Res.* **28:**1625–1634.

Cordell, B., DeNoto, F. M., Atkins, J. F., Gesteland, R. F., Bishop, J. M., and Goodman, H. M. (1980). The forms of tRNATrp found in avian sarcoma virus and uninfected chicken cells have structural identity but functional distinctions. *J. Biol. Chem.* **255:**9358–9368.

Costa-Mattioli, M., Svitkin, Y., and Sonenberg, N. (2004). La autoantigen is necessary for optimal function of the poliovirus and hepatitis C virus internal ribosome entry site *in vivo* and *in vitro*. *Mol. Cell. Biol.* **24:**6861–6870.

Costantino, D. A., Pfingsten, J. S., Rambo, R. P., and Kieft, J. S. (2008). tRNA-mRNA mimicry drives translation initiation from a viral IRES. *Nat. Struct. Mol. Biol.* **15:**57–64.

Cotton, S., Dufresne, P. J., Thivierge, K., Ide, C., and Fortin, M. G. (2006). The VPgPro protein of Turnip mosaic virus: *In vitro* inhibition of translation from a ribonuclease activity. *Virology* **351:**92–100.

Craig, A. W., Haghighat, A., Yu, A. T., and Sonenberg, N. (1998). Interaction of polyadenylate-binding protein with the eIF4G homologue PAIP enhances translation. *Nature* **392:**520–523.

Cuesta, R., Xi, Q., and Schneider, R. J. (2000). Adenovirus-specific translation by displacement of kinase Mnk1 from cap-initiation complex eIF4F. *EMBO J.* **19:**3465–3474.

Cullen, B. R. (1998). Retroviruses as model systems for the study of nuclear RNA export pathways. *Virology* **249:**203–210.

Curran, J., Boeck, R., and Kolakofsky, D. (1991). The Sendai virus P gene expresses both an essential protein and an inhibitor of RNA synthesis by shuffling modules via mRNA editing. *EMBO J.* **10:**3079–3085.

Curran, J., and Kolakofsky, D. (1988). Ribosomal initiation from an ACG codon in the Sendai virus P/C mRNA. *EMBO J.* **7:**245–251.

Danthinne, X., Seurinck, J., Meulewaeter, F., Van Montagu, M., and Cornelissen, M. (1993). The 3′ untranslated region of satellite tobacco necrosis virus RNA stimulates translation *in vitro*. *Mol. Cell. Biol.* **13:**3340–3349.

Daughenbaugh, K. F., Fraser, C. S., Hershey, J. W., and Hardy, M. E. (2003). The genome-linked protein VPg of the Norwalk virus binds eIF3, suggesting its role in translation initiation complex recruitment. *EMBO J.* **22:**2852–2859.

de la Luna, S., Fortes, P., Beloso, A., and Ortín, J. (1995). Influenza virus NS1 protein enhances the rate of translation initiation of viral mRNAs. *J. Virol.* **69:**2427–2433.

de la Torre, J. C. (2002). Molecular biology of Borna disease virus and persistence. *Front. Biosci.* **7:**D569–D579.

Demler, S. A., and de Zoeten, G. A. (1991). The nucleotide sequence and luteo-virus-like nature of RNA 1 of an aphid non-transmissible strain of pea enation mosaic virus. *J. Gen. Virol.* **72:**1819–1834.

Demler, S. A., Rucker, D. G., and de Zoeten, G. A. (1993). The chimeric nature of the genome of pea enation mosaic virus: The independent replication of RNA 2. *J. Gen. Virol.* **74:**1–14.

den Boon, J. A., Snijder, E. J., Chirnside, E. D., de Vries, A. A., Horzinek, M. C., and Spaan, W. J. (1991). Equine arteritis virus is not a togavirus but belongs to the coronaviruslike superfamily. *J. Virol.* **65:**2910–2920.

Deo, R. C., Groft, C. M., Rajashankar, K. R., and Burley, S. K. (2002). Recognition of the rotavirus mRNA 3′ consensus by an asymmetric NSP3 homodimer. *Cell* **108:**71–81.

Dever, T. E., Dar, A. C., and Sicheri, F. (2007). The eIF2α Kinases. *In* "Translational Control in Biology and Medicine" (M. B. Mathews, N. Sonenberg, and J. W. B. Hershey, eds.), pp. 319–344. CSHL Press, Cold Spring Harbor, NY.

Di, R., Dinesh-Kumar, S. P., and Miller, W. A. (1993). Translational frameshifting by barley yellow dwarf virus RNA (PAV serotype) in *Escherichia coli* and in eukaryotic cell-free extracts. *Mol. Plant Microbe Interact.* **6:**444–452.

Dinesh-Kumar, S. P., Brault, V., and Miller, W. A. (1992). Precise mapping and *in vitro* translation of a trifunctional subgenomic RNA of barley yellow dwarf virus. *Virology* **187:**711–722.

Dinman, J. D., Icho, T., and Wickner, R. B. (1991). A − 1 ribosomal frameshift in a double-stranded RNA virus of yeast forms a gag-pol fusion protein. *Proc. Natl. Acad. Sci. USA* **88:**174–178.

Domier, L. L., McCoppin, N. K., and D'Arcy, C. J. (2000). Sequence requirements for translation initiation of Rhopalosiphum padi virus ORF2. *Virology* **268:**264–271.

Doudna, J. A., and Sarnow, P. (2007). Translational initiation by viral internal ribosome entry sites. *In* "Translational Control in Biology and Medicine" (M. B. Mathews, N. Sonenberg, and J. W. B. Hershey, eds.), pp. 129–153. CSHL Press, Cold Spring Harbor, NY.

Dreher, T. W., and Miller, W. A. (2006). Translational control in positive strand RNA plant viruses. *Virology* **344:**185–197.

Drugeon, G., Jean-Jean, O., Frolova, L., Le Goff, X., Philippe, M., Kisselev, L., and Haenni, A. L. (1997). Eukaryotic release factor 1 (eRF1) abolishes readthrough and competes with suppressor tRNAs at all three termination codons in messenger RNA. *Nucleic Acids Res.* **25:**2254–2258.

Edgil, D., and Harris, E. (2006). End-to-end communication in the modulation of translation by mammalian RNA viruses. *Virus Res.* **119:**43–51.

Elgadi, M. M., Hayes, C. E., and Smiley, J. R. (1999). The herpes simplex virus vhs protein induces endoribonucleolytic cleavage of target RNAs in cell extracts. *J. Virol.* **73:**7153–7164.

Enami, K., Sato, T. A., Nakada, S., and Enami, M. (1994). Influenza virus NS1 protein stimulates translation of the M1 protein. *J. Virol.* **68:**1432–1437.

Etzerodt, M., Mikkelsen, T., Pedersen, F. S., Kjeldgaard, N. O., and Jorgensen, P. (1984). The nucleotide sequence of the Akv murine leukemia virus genome. *Virology* **134:**196–207.

Everly, D. N., Jr., Feng, P., Mian, I. S., and Read, G. S. (2002). mRNA degradation by the virion host shutoff (Vhs) protein of herpes simplex virus: Genetic and biochemical evidence that Vhs is a nuclease. *J. Virol.* **76:**8560–8571.

Fabian, M. R., and White, K. A. (2004). 5'-3' RNA-RNA interaction facilitates cap- and poly(A) tail-independent translation of tomato bushy stunt virus mRNA: A potential common mechanism for tombusviridae. *J. Biol. Chem.* **279:** 28862–28872.

Fabian, M. R., and White, K. A. (2006). Analysis of a 3'-translation enhancer in a tombusvirus: A dynamic model for RNA–RNA interactions of mRNA termini. *RNA* **12:**1304–1314.

Farabaugh, P. J. (2000). Translational frameshifting: Implications for the mechanism of translational frame maintenance. *Prog. Nucleic Acid Res. Mol. Biol.* **64:**131–170.

Feigenblum, D., and Schneider, R. J. (1993). Modification of eukaryotic initiation factor 4F during infection by influenza virus. *J. Virol.* **67:**3027–3035.

Filichkin, S. A., Lister, R. M., McGrath, P. F., and Young, M. J. (1994). *In vivo* expression and mutational analysis of the barley yellow dwarf virus readthrough gene. *Virology* **205:**290–299.

Flynn, A., and Proud, C. G. (1995). Serine 209, not serine 53, is the major site of phosphorylation in initiation factor eIF-4E in serum-treated Chinese hamster ovary cells. *J. Biol. Chem.* **270:**21684–21688.

Frank, J., Gao, H., Sengupta, J., Gao, N., and Taylor, D. J. (2007). The process of mRNA-tRNA translocation. *Proc. Natl. Acad. Sci. USA* **104:**19671–19678.

Fukuhara, N., Huang, C., Kiyotani, K., Yoshida, T., and Sakaguchi, T. (2002). Mutational analysis of the Sendai virus V protein: Importance of the conserved residues for Zn binding, virus pathogenesis, and efficient RNA editing. *Virology* **299:**172–181.

Fütterer, J., and Hohn, T. (1991). Translation of a polycistronic mRNA in the presence of the cauliflower mosaic virus transactivator protein. *EMBO J.* **10:**3887–3896.

Fütterer, J., and Hohn, T. (1996). Translation in plants—Rules and exceptions. *Plant Mol. Biol.* **32:**159–189.

Fütterer, J., Potrykus, I., Bao, Y., Li, L., Burns, T. M., Hull, R., and Hohn, T. (1996). Position-dependent ATT initiation during plant pararetrovirus rice tungro bacilliform virus translation. *J. Virol.* **70:**2999–3010.

Fütterer, J., Rothnie, H. M., Hohn, T., and Potrykus, I. (1997). Rice tungro bacilliform virus open reading frames II and III are translated from polycistronic pregenomic RNA by leaky scanning. *J. Virol.* **71:**7984–7989.

Gale, M., Jr., Tan, S. L., and Katze, M. G. (2000). Translational control of viral gene expression in eukaryotes. *Microbiol. Mol. Biol. Rev.* **64:**239–280.

Gallie, D. R. (1998). A tale of two termini: A functional interaction between the termini of an mRNA is a prerequisite for efficient translation initiation. *Gene* **216:**1–11.

Gamarnik, A. V., and Andino, R. (1998). Switch from translation to RNA replication in a positive-stranded RNA virus. *Genes Dev.* **12:**2293–2304.

Garcia, M. A., Meurs, E. F., and Esteban, M. (2007). The dsRNA protein kinase PKR: Virus and cell control. *Biochimie* **89:**799–811.

Garcin, D., and Kolakofsky, D. (1990). A novel mechanism for the initiation of tacaribe arenavirus genome replication. *J. Virol.* **64:**6196–6203.

Garcin, D., Lezzi, M., Dobbs, M., Elliott, R. M., Schmaljohn, C., Kang, C. Y., and Kolakofsky, D. (1995). The 5′ ends of Hantaan virus (Bunyaviridae) RNAs suggest a prime-and-realign mechanism for the initiation of RNA synthesis. *J. Virol.* **69:**5754–5762.

Garfinkel, M. S., and Katze, M. G. (1992). Translational control by influenza virus. Selective and cap-dependent translation of viral mRNAs in infected cells. *J. Biol. Chem.* **267:**9383–9390.

Gazo, B. M., Murphy, P., Gatchel, J. R., and Browning, K. S. (2004). A novel interaction of Cap-binding protein complexes eukaryotic initiation factor (eIF) 4F and eIF(iso)4F with a region in the 3′-untranslated region of satellite tobacco necrosis virus. *J. Biol. Chem.* **279:**13584–13592.

Ge, Z., Hiruki, C., and Roy, K. L. (1993). Nucleotide sequence of sweet clover necrotic mosaic dianthovirus RNA-1. *Virus Res.* **28:**113–124.

Geller, A. I., and Rich, A. (1980). A UGA termination suppression tRNATrp active in rabbit reticulocytes. *Nature* **283:**41–46.

German-Retana, S., Walter, J., Doublet, B., Roudet-Tavert, G., Nicaise, V., Lecampion, C., Houvenaghel, M. C., Robaglia, C., Michon, T., and Le Gall, O. (2008). Mutational analysis of a plant cap-binding protein eIF4E reveals key amino-acids involved in biochemical functions and potyvirus infection. *J. Virol.* **92:**7601–7612.

Gingras, A. C., Svitkin, Y., Belsham, G. J., Pause, A., and Sonenberg, N. (1996). Activation of the translational suppressor 4E-BP1 following infection with encephalomyocarditis virus and poliovirus. *Proc. Natl. Acad. Sci. USA* **93:**5578–5583.

Goff, S. P. (2004). Genetic reprogramming by retroviruses: Enhanced suppression of translational termination. *Cell Cycle* **3:**123–125.

Goodfellow, I., Chaudhry, Y., Gioldasi, I., Gerondopoulos, A., Natoni, A., Labrie, L., Laliberte, J. F., and Roberts, R. (2005). Calicivirus translation initiation requires an interaction between VPg and eIF4E. *EMBO Rep.* **6:**968–972.

Gosert, R., Chang, K. H., Rijnbrand, R., Yi, M., Sangar, D. V., and Lemon, S. M. (2000). Transient expression of cellular polypyrimidine-tract binding protein stimulates cap-independent translation directed by both picornaviral and flaviviral internal ribosome entry sites *in vivo*. *Mol. Cell. Biol.* **20:**1583–1595.

Gradi, A., Foeger, N., Strong, R., Svitkin, Y. V., Sonenberg, N., Skern, T., and Belsham, G. J. (2004). Cleavage of eukaryotic translation initiation factor 4GII within foot-and-mouth disease virus-infected cells: Identification of the L-protease cleavage site *in vitro*. *J. Virol.* **78:**3271–3278.

Gradi, A., Svitkin, Y. V., Imataka, H., and Sonenberg, N. (1998). Proteolysis of human eukaryotic translation initiation factor eIF4GII, but not eIF4GI, coincides with the shutoff of host protein synthesis after poliovirus infection. *Proc. Natl. Acad. Sci. USA* **95:**11089–11094.

Gramstat, A., Prüfer, D., and Rohde, W. (1994). The nucleic acid-binding zinc finger protein of potato virus M is translated by internal initiation as well as by ribosomal frameshifting involving a shifty stop codon and a novel mechanism of P-site slippage. *Nucleic Acids Res.* **22**:3911–3917.

Grieco, F., Burgyan, J., and Russo, M. (1989). The nucleotide sequence of cymbidium ringspot virus RNA. *Nucleic Acids Res.* **17**:6383.

Gudima, S., Wu, S. Y., Chiang, C. M., Moraleda, G., and Taylor, J. (2000). Origin of hepatitis delta virus mRNA. *J. Virol.* **74**:7204–7210.

Guilley, H., Carrington, J. C., Balazs, E., Jonard, G., Richards, K., and Morris, T. J. (1985). Nucleotide sequence and genome organization of carnation mottle virus RNA. *Nucleic Acids Res.* **13**:6663–6677.

Guo, L., Allen, E., and Miller, W. A. (2000). Structure and function of a cap-independent translation element that functions in either the 3′ or the 5′ untranslated region. *RNA* **6**:1808–1820.

Guo, L., Allen, E. M., and Miller, W. A. (2001). Base-pairing between untranslated regions facilitates translation of uncapped, nonpolyadenylated viral RNA. *Mol. Cell* **7**:1103–1109.

Gupta, K. C., and Patwardhan, S. (1988). ACG, the initiator codon for a Sendai virus protein. *J. Biol. Chem.* **263**:8553–8556.

Haghighat, A., Mader, S., Pause, A., and Sonenberg, N. (1995). Repression of cap-dependent translation by 4E-binding protein 1: Competition with p220 for binding to eukaryotic initiation factor-4E. *EMBO J.* **14**:5701–5709.

Hamilton, W. D., Boccara, M., Robinson, D. J., and Baulcombe, D. C. (1987). The complete nucleotide sequence of tobacco rattle virus RNA-1. *J. Gen. Virol.* **68**:2563–2575.

Hansen, L. J., Chalker, D. L., Orlinsky, K. J., and Sandmeyer, S. B. (1992). Ty3 GAG3 and POL3 genes encode the components of intracellular particles. *J. Virol.* **66**:1414–1424.

Hearne, P. Q., Knorr, D. A., Hillman, B. I., and Morris, T. J. (1990). The complete genome structure and synthesis of infectious RNA from clones of tomato bushy stunt virus. *Virology* **177**:141–151.

Hentze, M. W., Gebauer, F., and Preiss, T. (2007). *cis*-Regulatory sequences and trans-acting factors in translational control. *In* "Translational Control in Biology and Medicine" (M. B. Mathews, N. Sonenberg, and J. W. B. Hershey, eds.), pp. 269–295. CSHL Press, Cold Spring Harbor, NY.

Herbert, T. P., Brierley, I., and Brown, T. D. (1997). Identification of a protein linked to the genomic and subgenomic mRNAs of feline calicivirus and its role in translation. *J. Gen. Virol.* **78**:1033–1040.

Herold, J., and Andino, R. (2001). Poliovirus RNA replication requires genome circularization through a protein-protein bridge. *Mol. Cell* **7**:581–591.

Herr, W. (1984). Nucleotide sequence of AKV murine leukemia virus. *J. Virol.* **49**:471–478.

Herz, C., Stavnezer, E., Krug, R., and Gurney, T., Jr. (1981). Influenza virus, an RNA virus, synthesizes its messenger RNA in the nucleus of infected cells. *Cell* **26**:391–400.

Herzog, E., Guilley, H., and Fritsch, C. (1995). Translation of the second gene of peanut clump virus RNA 2 occurs by leaky scanning *in vitro*. *Virology* **208**:215–225.

Herzog, E., Guilley, H., Manohar, S. K., Dollet, M., Richards, K., Fritsch, C., and Jonard, G. (1994). Complete nucleotide sequence of peanut clump virus RNA 1 and relationships with other fungus-transmitted rod-shaped viruses. *J. Gen. Virol.* **75**:3147–3155.

Hilton, A., Mizzen, L., MacIntyre, G., Cheley, S., and Anderson, R. (1986). Translational control in murine hepatitis virus infection. *J. Gen. Virol.* **67:**923–932.

Hinnebusch, A. G. (2005). Translational regulation of GCN4 and the general amino acid control of yeast. *Annu. Rev. Microbiol.* **59:**407–450.

Hinnebusch, A. G. (2006). eIF3: A versatile scaffold for translation initiation complexes. *Trends Biochem. Sci.* **31:**553–562.

Holzschu, D. L., Martineau, D., Fodor, S. K., Vogt, V. M., Bowser, P. R., and Casey, J. W. (1995). Nucleotide sequence and protein analysis of a complex piscine retrovirus, walleye dermal sarcoma virus. *J. Virol.* **69:**5320–5331.

Honda, M., Brown, E. A., and Lemon, S. M. (1996). Stability of a stem loop involving the initiator AUG controls the efficiency of internal initiation of translation on hepatitis C virus RNA. *RNA* **2:**955–968.

Huiet, L., Feldstein, P. A., Tsai, J. H., and Falk, B. W. (1993). The maize stripe virus major noncapsid protein messenger RNA transcripts contain heterogeneous leader sequences at their 5′ termini. *Virology* **197:**808–812.

Hunt, S. L., and Jackson, R. J. (1999). Polypyrimidine-tract binding protein (PTB) is necessary, but not sufficient, for efficient internal initiation of translation of human rhinovirus-2 RNA. *RNA* **5:**344–359.

Imataka, H., Gradi, A., and Sonenberg, N. (1998). A newly identified N-terminal amino acid sequence of human eIF4G binds poly(A)-binding protein and functions in poly(A)-dependent translation. *EMBO J.* **17:**7480–7489.

Ishikawa, M., Meshi, T., Motoyoshi, F., Takamatsu, N., and Okada, Y. (1986). *In vitro* mutagenesis of the putative replicase genes of tobacco mosaic virus. *Nucleic Acids Res.* **14:**8291–8305.

Jaag, H. M., Kawchuk, L., Rohde, W., Fischer, R., Emans, N., and Prüfer, D. (2003). An unusual internal ribosomal entry site of inverted symmetry directs expression of a potato leafroll polerovirus replication-associated protein. *Proc. Natl. Acad. Sci. USA* **100:**8939–8944.

Jacks, T., Townsley, K., Varmus, H. E., and Majors, J. (1987). Two efficient ribosomal frameshifting events are required for synthesis of mouse mammary tumor virus gag-related polyproteins. *Proc. Natl. Acad. Sci. USA* **84:**4298–4302.

Jackson, R. J. (2005). Alternative mechanisms of initiating translation of mammalian mRNAs. *Biochem. Soc. Trans.* **33:**1231–1241.

Jackson, R. J., and Kaminski, A. (1995). Internal initiation of translation in eukaryotes: The picornavirus paradigm and beyond. *RNA* **1:**985–1000.

James, D., Varga, A., and Croft, H. (2007). Analysis of the complete genome of peach chlorotic mottle virus: Identification of non-AUG start codons, *in vitro* coat protein expression, and elucidation of serological cross-reactions. *Arch. Virol.* **152:**2207–2215.

Jan, E. (2006). Divergent IRES elements in invertebrates. *Virus Res.* **119:**16–28.

Jan, E., Kinzy, T. G., and Sarnow, P. (2003). Divergent tRNA-like element supports initiation, elongation, and termination of protein biosynthesis. *Proc. Natl. Acad. Sci. USA* **100:**15410–15415.

Jan, E., and Sarnow, P. (2002). Factorless ribosome assembly on the internal ribosome entry site of cricket paralysis virus. *J. Mol. Biol.* **324:**889–902.

Jang, S. K., Krausslich, H. G., Nicklin, M. J., Duke, G. M., Palmenberg, A. C., and Wimmer, E. (1988). A segment of the 5′ nontranslated region of encephalomyocarditis virus RNA directs internal entry of ribosomes during *in vitro* translation. *J. Virol.* **62:**2636–2643.

Janzen, D. M., Frolova, L., and Geballe, A. P. (2002). Inhibition of translation termination mediated by an interaction of eukaryotic release factor 1 with a nascent peptidyl-tRNA. *Mol. Cell. Biol.* **22:**8562–8570.

Joachims, M., Van Breugel, P. C., and Lloyd, R. E. (1999). Cleavage of poly(A)-binding protein by enterovirus proteases concurrent with inhibition of translation *in vitro*. *J. Virol.* **73:**718–727.

Jordan, I., and Lipkin, W. I. (2001). Borna disease virus. *Rev. Med. Virol.* **11:**37–57.

Joshi, B., Cai, A. L., Keiper, B. D., Minich, W. B., Mendez, R., Beach, C. M., Stepinski, J., Stolarski, R., Darzynkiewicz, E., and Rhoads, R. E. (1995). Phosphorylation of eukaryotic protein synthesis initiation factor 4E at Ser-209. *J. Biol. Chem.* **270:**14597–145603.

Kamitani, W., Narayanan, K., Huang, C., Lokugamage, K., Ikegami, T., Ito, N., Kubo, H., and Makino, S. (2006). Severe acute respiratory syndrome coronavirus nsp1 protein suppresses host gene expression by promoting host mRNA degradation. *Proc. Natl. Acad. Sci. USA* **103:**12885–12890.

Kang, B. C., Yeam, I., and Jahn, M. M. (2005). Genetics of plant virus resistance. *Annu. Rev. Phytopathol.* **43:**581–621.

Karamysheva, Z. N., Karamyshev, A. L., Ito, K., Yokogawa, T., Nishikawa, K., Nakamura, Y., and Matsufuji, S. (2003). Antizyme frameshifting as a functional probe of eukaryotic translational termination. *Nucleic Acids Res.* **31:**5949–5956.

Karasev, A. V., Boyko, V. P., Gowda, S., Nikolaeva, O. V., Hilf, M. E., Koonin, E. V., Niblett, C. L., Cline, K., Gumpf, D. J., Lee, R. F., Garnsey, S. M., Lewandowski, D. J., *et al.* (1995). Complete sequence of the citrus tristeza virus RNA genome. *Virology* **208:**511–520.

Kato, A., Cortese-Grogan, C., Moyer, S. A., Sugahara, F., Sakaguchi, T., Kubota, T., Otsuki, N., Kohase, M., Tashiro, M., and Nagai, Y. (2004). Characterization of the amino acid residues of sendai virus C protein that are critically involved in its interferon antagonism and RNA synthesis down-regulation. *J. Virol.* **78:**7443–7454.

Kato, K., Kawaguchi, Y., Tanaka, M., Igarashi, M., Yokoyama, A., Matsuda, G., Kanamori, M., Nakajima, K., Nishimura, Y., Shimojima, M., Phung, H. T., Takahashi, E., *et al.* (2001). Epstein–Barr virus-encoded protein kinase BGLF4 mediates hyperphosphorylation of cellular elongation factor 1delta (EF-1delta): EF-1delta is universally modified by conserved protein kinases of herpesviruses in mammalian cells. *J. Gen. Virol.* **82:**1457–1463.

Katsafanas, G. C., and Moss, B. (2007). Colocalization of transcription and translation within cytoplasmic poxvirus factories coordinates viral expression and subjugates host functions. *Cell Host Microbe* **2:**221–228.

Katze, M. G., DeCorato, D., and Krug, R. M. (1986). Cellular mRNA translation is blocked at both initiation and elongation after infection by influenza virus or adenovirus. *J. Virol.* **60:**1027–1039.

Katze, M. G., and Krug, R. M. (1990). Translational control in influenza virus-infected cells. *Enzyme* **44:**265–277.

Kawaguchi, Y., Kato, K., Tanaka, M., Kanamori, M., Nishiyama, Y., and Yamanashi, Y. (2003). Conserved protein kinases encoded by herpesviruses and cellular protein kinase cdc2 target the same phosphorylation site in eukaryotic elongation factor 1delta. *J. Virol.* **77:**2359–2368.

Kawaguchi, Y., Matsumura, T., Roizman, B., and Hirai, K. (1999). Cellular elongation factor 1delta is modified in cells infected with representative alpha-, beta-, or gammaherpesviruses. *J. Virol.* **73:**4456–4460.

Kean, K. M., Michel, Y. M., Malnou, C. E., Paulous, S., and Borman, A. M. (2001). Roles and mechanisms of mRNA 5′–3′ end cross-talk in translation initiation on animal virus RNAs. *Rec. Res. Dev. Virol.* **3:**165–176.

Kerekatte, V., Keiper, B. D., Badorff, C., Cai, A., Knowlton, K. U., and Rhoads, R. E. (1999). Cleavage of Poly(A)-binding protein by coxsackievirus 2A protease *in vitro* and *in vivo*: Another mechanism for host protein synthesis shutoff? *J. Virol.* **73**:709–717.

Khan, M. A., Miyoshi, H., Gallie, D. R., and Goss, D. J. (2008). Potyvirus genome-linked protein, VPg, directly affects wheat germ *in vitro* translation: Interactions with translation initiation factors eIF4F and eIFiso4F. *J. Biol. Chem.* **283**:1340–1349.

Kim, K. H., and Lommel, S. A. (1994). Identification and analysis of the site of − 1 ribosomal frameshifting in red clover necrotic mosaic virus. *Virology* **200**:574–582.

Kim, Y. G., Maas, S., and Rich, A. (2001). Comparative mutational analysis of cis-acting RNA signals for translational frameshifting in HIV-1 and HTLV-2. *Nucleic Acids Res.* **29**:1125–1131.

Klaassen, V. A., Boeshore, M. L., Koonin, E. V., Tian, T., and Falk, B. W. (1995). Genome structure and phylogenetic analysis of lettuce infectious yellows virus, a whitefly-transmitted, bipartite closterovirus. *Virology* **208**:99–110.

Kneller, E. L., Rakotondrafara, A. M., and Miller, W. A. (2006). Cap-independent translation of plant viral RNAs. *Virus Res.* **119**:63–75.

Kobayashi, T., Watanabe, M., Kamitani, W., Tomonaga, K., and Ikuta, K. (2000). Translation initiation of a bicistronic mRNA of Borna disease virus: A 16-kDa phosphoprotein is initiated at an internal start codon. *Virology* **277**:296–305.

Koenig, R., Commandeur, U., Loss, S., Beier, C., Kaufmann, A., and Lesemann, D. E. (1997). Beet soil-borne virus RNA 2: Similarities and dissimilarities to the coat protein gene-carrying RNAs of other furoviruses. *J. Gen. Virol.* **78**:469–477.

Koenig, R., and Loss, S. (1997). Beet soil-borne virus RNA 1: Genetic analysis enabled by a starting sequence generated with primers to highly conserved helicase-encoding domains. *J. Gen. Virol.* **78**:3161–3165.

Koenig, R., Pleij, C. W., Beier, C., and Commandeur, U. (1998). Genome properties of beet virus Q, a new furo-like virus from sugarbeet, determined from unpurified virus. *J. Gen. Virol.* **79**:2027–2036.

Koh, D. C., Liu, D. X., and Wong, S. M. (2002). A six-nucleotide segment within the 3′ untranslated region of hibiscus chlorotic ringspot virus plays an essential role in translational enhancement. *J. Virol.* **76**:1144–1153.

Koh, D. C., Wong, S. M., and Liu, D. X. (2003). Synergism of the 3′-untranslated region and an internal ribosome entry site differentially enhances the translation of a plant virus coat protein. *J. Biol. Chem.* **278**:20565–20573.

Komarova, A. V., Brocard, M., and Kean, K. M. (2006). The case for mRNA 5′ and 3′ end cross talk during translation in a eukaryotic cell. *Prog. Nucleic Acid Res. Mol. Biol.* **81**:331–367.

Kononenko, A. V., Mitkevich, V. A., Dubovaya, V. I., Kolosov, P. M., Makarov, A. A., and Kisselev, L. L. (2008). Role of the individual domains of translation termination factor eRF1 in GTP binding to eRF3. *Proteins* **70**:388–393.

Kovacs, G. R., Guarino, L. A., Graham, B. L., and Summers, M. D. (1991). Identification of spliced baculovirus RNAs expressed late in infection. *Virology* **185**:633–643.

Kozak, M. (1986). Point mutations define a sequence flanking the AUG initiator codon that modulates translation by eukaryotic ribosomes. *Cell* **44**:283–292.

Kozak, M. (1991). Structural features in eukaryotic mRNAs that modulate the initiation of translation. *J. Biol. Chem.* **266**:19867–19870.

Kozak, M., and Shatkin, A. J. (1978). Migration of 40 S ribosomal subunits on messenger RNA in the presence of edeine. *J. Biol. Chem.* **253**:6568–6577.

Krab, I. M., Caldwell, C., Gallie, D. R., and Bol, J. F. (2005). Coat protein enhances translational efficiency of Alfalfa mosaic virus RNAs and interacts with the eIF4G component of initiation factor eIF4F. *J. Gen. Virol.* **86:**1841–1849.

Krishnamoorthy, T., Pavitt, G. D., Zhang, F., Dever, T. E., and Hinnebusch, A. G. (2001). Tight binding of the phosphorylated alpha subunit of initiation factor 2 (eIF2alpha) to the regulatory subunits of guanine nucleotide exchange factor eIF2B is required for inhibition of translation initiation. *Mol. Cell. Biol.* **21:**5018–5030.

Krug, R. M., Broni, B. A., and Bouloy, M. (1979). Are the 5′ ends of influenza viral mRNAs synthesized *in vivo* donated by host mRNAs? *Cell* **18:**329–334.

Kudchodkar, S. B., Yu, Y., Maguire, T. G., and Alwine, J. C. (2004). Human cytomegalovirus infection induces rapamycin-insensitive phosphorylation of downstream effectors of mTOR kinase. *J. Virol.* **78:**11030–11039.

Kudchodkar, S. B., Yu, Y., Maguire, T. G., and Alwine, J. C. (2006). Human cytomegalovirus infection alters the substrate specificities and rapamycin sensitivities of raptor- and rictor-containing complexes. *Proc. Natl. Acad. Sci. USA* **103:**14182–14187.

Kujawa, A. B., Drugeon, G., Hulanicka, D., and Haenni, A. L. (1993). Structural requirements for efficient translational frameshifting in the synthesis of the putative viral RNA-dependent RNA polymerase of potato leafroll virus. *Nucleic Acids Res.* **21:**2165–2171.

Kuyumcu-Martinez, M., Belliot, G., Sosnovtsev, S. V., Chang, K. O., Green, K. Y., and Lloyd, R. E. (2004a). Calicivirus 3C-like proteinase inhibits cellular translation by cleavage of poly(A)-binding protein. *J. Virol.* **78:**8172–8182.

Kuyumcu-Martinez, N. M., Joachims, M., and Lloyd, R. E. (2002). Efficient cleavage of ribosome-associated poly(A)-binding protein by enterovirus 3C protease. *J. Virol.* **76:**2062–2074.

Kuyumcu-Martinez, N. M., Van Eden, M. E., Younan, P., and Lloyd, R. E. (2004b). Cleavage of poly(A)-binding protein by poliovirus 3C protease inhibits host cell translation: A novel mechanism for host translation shutoff. *Mol. Cell. Biol.* **24:**1779–1790.

Lai, M. M., and Cavanagh, D. (1997). The molecular biology of coronaviruses. *Adv. Virus Res.* **48:**1–100.

Lamb, R. A., and Krug, R. M. (2001). Orthomyxoviridae: The viruses and their replication. *In* "Fields Virology" (D. M. Knipe and P. W. Howley, eds.), pp. 1487–1531. Lippincott Williams & Wilkens, Philadelphia, PA.

Lamphear, B. J., Yan, R., Yang, F., Waters, D., Liebig, H. D., Klump, H., Kuechler, E., Skern, T., and Rhoads, R. E. (1993). Mapping the cleavage site in protein synthesis initiation factor eIF-4 gamma of the 2A proteases from human Coxsackievirus and rhinovirus. *J. Biol. Chem.* **268:**19200–19203.

Latorre, P., Kolakofsky, D., and Curran, J. (1998). Sendai virus Y proteins are initiated by a ribosomal shunt. *Mol. Cell. Biol.* **18:**5021–5031.

Le, H., Tanguay, R. L., Balasta, M. L., Wei, C. C., Browning, K. S., Metz, A. M., Goss, D. J., and Gallie, D. R. (1997). Translation initiation factors eIF-iso4G and eIF-4B interact with the poly(A)-binding protein and increase its RNA binding activity. *J. Biol. Chem.* **272:**16247–16255.

Leh, V., Yot, P., and Keller, M. (2000). The cauliflower mosaic virus translational transactivator interacts with the 60S ribosomal subunit protein L18 of *Arabidopsis thaliana*. *Virology* **266:**1–7.

Leonard, S., Viel, C., Beauchemin, C., Daigneault, N., Fortin, M. G., and Laliberté, J. F. (2004). Interaction of VPg-Pro of turnip mosaic virus with the translation initiation factor 4E and the poly(A)-binding protein in planta. *J. Gen. Virol.* **85:**1055–1063.

Leong, W. F., Tan, H. C., Ooi, E. E., Koh, D. R., and Chow, V. T. (2005). Microarray and real-time RT-PCR analyses of differential human gene expression patterns induced by severe acute respiratory syndrome (SARS) coronavirus infection of Vero cells. *Microbes Infect.* **7:**248–259.

Levis, C., and Astier-Manifacier, S. (1993). The 5′ untranslated region of PVY RNA, even located in an internal position, enables initiation of translation. *Virus Genes* **7:**367–379.

Lewis, T. L., and Matsui, S. M. (1996). Astrovirus ribosomal frameshifting in an infection-transfection transient expression system. *J. Virol.* **70:**2869–2875.

Lin, T. A., Kong, X., Haystead, T. A., Pause, A., Belsham, G., Sonenberg, N., and Lawrence, J. C., Jr. (1994). PHAS-I as a link between mitogen-activated protein kinase and translation initiation. *Science* **266:**653–656.

Liston, P., and Briedis, D. J. (1995). Ribosomal frameshifting during translation of measles virus P protein mRNA is capable of directing synthesis of a unique protein. *J. Virol.* **69:**6742–6750.

Lloyd, R. E. (2006). Translational control by viral proteinases. *Virus Res.* **119:**76–88.

Luo, G. X., Luytjes, W., Enami, M., and Palese, P. (1991). The polyadenylation signal of influenza virus RNA involves a stretch of uridines followed by the RNA duplex of the panhandle structure. *J. Virol.* **65:**2861–2867.

Luttermann, C., and Meyers, G. (2007). A bipartite sequence motif induces translation reinitiation in feline calicivirus RNA. *J. Biol. Chem.* **282:**7056–7065.

MacFarlane, S. A., Taylor, S. C., King, D. I., Hughes, G., and Davies, J. W. (1989). Pea early browning virus RNA1 encodes four polypeptides including a putative zinc-finger protein. *Nucleic Acids Res.* **17:**2245–2260.

Mahapatra, M., Parida, S., Egziabher, B. G., Diallo, A., and Barrett, T. (2003). Sequence analysis of the phosphoprotein gene of peste des petits ruminants (PPR) virus: Editing of the gene transcript. *Virus Res.* **96:**85–98.

Maia, I. G., Séron, K., Haenni, A. L., and Bernardi, F. (1996). Gene expression from viral RNA genomes. *Plant Mol. Biol.* **32:**367–391.

Mäkinen, K., Naess, V., Tamm, T., Truve, E., Aaspollu, A., and Saarma, M. (1995). The putative replicase of the cocksfoot mottle sobemovirus is translated as a part of the polyprotein by − 1 ribosomal frameshift. *Virology* **207:**566–571.

Makkinje, A., Xiong, H., Li, M., and Damuni, Z. (1995). Phosphorylation of eukaryotic protein synthesis initiation factor 4E by insulin-stimulated protamine kinase. *J. Biol. Chem.* **270:**14824–14828.

Marcotrigiano, J., Gingras, A. C., Sonenberg, N., and Burley, S. K. (1999). Cap-dependent translation initiation in eukaryotes is regulated by a molecular mimic of eIF4G. *Mol. Cell* **3:**707–716.

Martínez-Salas, E., and Fernández-Miragall, O. (2004). Picornavirus IRES: Structure function relationship. *Curr. Pharm. Des.* **10:**3757–3767.

Martínez-Salas, E., Ramos, R., Lafuente, E., and López de Quinto, S. (2001). Functional interactions in internal translation initiation directed by viral and cellular IRES elements. *J. Gen. Virol.* **82:**973–984.

Matsuda, D., and Dreher, T. W. (2007). Cap- and initiator tRNA-dependent initiation of TYMV polyprotein synthesis by ribosomes: Evaluation of the Trojan horse model for TYMV RNA translation. *RNA* **13:**129–137.

Mazumder, B., Seshadri, V., and Fox, P. L. (2003). Translational control by the 3′-UTR: The ends specify the means. *Trends Biochem. Sci.* **28:**91–98.

Merrick, W. C., and Anderson, W. F. (1975). Purification and characterization of homogeneous protein synthesis initiation factor M1 from rabbit reticulocytes. *J. Biol. Chem.* **250:**1197–1206.

Meulewaeter, F., Danthinne, X., Van Montagu, M., and Cornelissen, M. (1998). 5'- and 3'-sequences of satellite tobacco necrosis virus RNA promoting translation in tobacco. *Plant J.* **14:**169–176.

Meulewaeter, F., Seurinck, J., and Van Emmelo, J. (1990). Genome structure of tobacco necrosis virus strain A. *Virology* **177:**699–709.

Meulewaeter, F., van Lipzig, R., Gultyaev, A. P., Pleij, C. W., Van Damme, D., Cornelissen, M., and van Eldik, G. (2004). Conservation of RNA structures enables TNV and BYDV 5' and 3' elements to cooperate synergistically in cap-independent translation. *Nucleic Acids Res.* **32:**1721–1730.

Meyers, G. (2003). Translation of the minor capsid protein of a calicivirus is initiated by a novel termination-dependent reinitiation mechanism. *J. Biol. Chem.* **278:**34051–34060.

Meyers, G. (2007). Characterization of the sequence element directing translation reinitiation in RNA of the calicivirus rabbit hemorrhagic disease virus. *J. Virol.* **81:**9623–9632.

Michel, Y. M., Poncet, D., Piron, M., Kean, K. M., and Borman, A. M. (2000). Cap-poly(A) synergy in mammalian cell-free extracts. Investigation of the requirements for poly(A)-mediated stimulation of translation initiation. *J. Biol. Chem.* **275:**32268–32276.

Miller, W. A., and Koev, G. (2000). Synthesis of subgenomic RNAs by positive-strand RNA viruses. *Virology* **273:**1–8.

Miller, W. A., Wang, Z., and Treder, K. (2007). The amazing diversity of cap-independent translation elements in the 3'-untranslated regions of plant viral RNAs. *Biochem. Soc. Trans.* **35:**1629–1633.

Miller, W. A., Waterhouse, P. M., and Gerlach, W. L. (1988). Sequence and organization of barley yellow dwarf virus genomic RNA. *Nucleic Acids Res.* **16:**6097–6111.

Miller, W. A., and White, K. A. (2006). Long-distance RNA-RNA interactions in plant virus gene expression and replication. *Annu. Rev. Phytopathol.* **44:**447–467.

Miyoshi, H., Suehiro, N., Tomoo, K., Muto, S., Takahashi, T., Tsukamoto, T., Ohmori, T., and Natsuaki, T. (2006). Binding analyses for the interaction between plant virus genome-linked protein (VPg) and plant translational initiation factors. *Biochimie* **88:**329–340.

Mohr, I. (2006). Phosphorylation and dephosphorylation events that regulate viral mRNA translation. *Virus Res.* **119:**89–99.

Mohr, I. J., Pe'ery, T., and Mathews, M. B. (2007). Protein synthesis and translational control during viral infection. *In* "Translational Control in Biology and Medicine" (M. B. Mathews, N. Sonenberg, and J. W. B. Hershey, eds.), pp. 545–599. CSHL Press, Cold Spring Harbor, NY.

Molina, S., Sanz, M. A., Madan, V., Ventoso, I., Castello, A., and Carrasco, L. (2007). Differential inhibition of cellular and Sindbis virus translation by brefeldin A. *Virology* **363:**430–436.

Montero, H., Arias, C. F., and López, S. (2006). Rotavirus nonstructural protein NSP3 is not required for viral protein synthesis. *J. Virol.* **80:**9031–9038.

Montero, H., Rojas, M., Arias, C. F., and López, S. (2008). Rotavirus infection induces the phosphorylation of eIF2{alpha} but prevents the formation of stress granules. *J. Virol.* **82:**1496–1504.

Moody, C. A., Scott, R. S., Amirghahari, N., Nathan, C. A., Young, L. S., Dawson, C. W., and Sixbey, J. W. (2005). Modulation of the cell growth regulator mTOR by Epstein–Barr virus-encoded LMP2A. *J. Virol.* **79:**5499–5506.

Munger, K., Baldwin, A., Edwards, K. M., Hayakawa, H., Nguyen, C. L., Owens, M., Grace, M., and Huh, K. (2004). Mechanisms of human papillomavirus-induced oncogenesis. *J. Virol.* **78:**11451–11460.

Nagai, Y. (1999). Paramyxovirus replication and pathogenesis. Reverse genetics transforms understanding. *Rev. Med. Virol.* **9:**83–99.

Nam, S. H., Copeland, T. D., Hatanaka, M., and Oroszlan, S. (1993). Characterization of ribosomal frameshifting for expression of pol gene products of human T-cell leukemia virus type I. *J. Virol.* **67:**196–203.

Napthine, S., Vidakovic, M., Girnary, R., Namy, O., and Brierley, I. (2003). Prokaryotic-style frameshifting in a plant translation system: Conservation of an unusual single-tRNA slippage event. *EMBO J.* **22:**3941–3950.

Neeleman, L., Olsthoorn, R. C., Linthorst, H. J., and Bol, J. F. (2001). Translation of a nonpolyadenylated viral RNA is enhanced by binding of viral coat protein or polyadenylation of the RNA. *Proc. Natl. Acad. Sci. USA* **98:**14286–14291.

Niesbach-Klosgen, U., Guilley, H., Jonard, G., and Richards, K. (1990). Immunodetection *in vivo* of beet necrotic yellow vein virus-encoded proteins. *Virology* **178:**52–61.

Nutter, R. C., Scheets, K., Panganiban, L. C., and Lommel, S. A. (1989). The complete nucleotide sequence of the maize chlorotic mottle virus genome. *Nucleic Acids Res.* **17:**3163–3177.

Oh, K. J., Kalinina, A., Park, N. H., and Bagchi, S. (2006). Deregulation of eIF4E: 4E-BP1 in differentiated human papillomavirus-containing cells leads to high levels of expression of the E7 oncoprotein. *J. Virol.* **80:**7079–7088.

Ohlmann, T., Prevot, D., Decimo, D., Roux, F., Garin, J., Morley, S. J., and Darlix, J. L. (2002). *In vitro* cleavage of eIF4GI but not eIF4GII by HIV-1 protease and its effects on translation in the rabbit reticulocyte lysate system. *J. Mol. Biol.* **318:**9–20.

Oosterom-Dragon, E. A., and Ginsberg, H. S. (1980). Purification and preliminary immunological characterization of the type 5 adenovirus, nonstructural 100,000-dalton protein. *J. Virol.* **33:**1203–1207.

Orlova, M., Yueh, A., Leung, J., and Goff, S. P. (2003). Reverse transcriptase of Moloney murine leukemia virus binds to eukaryotic release factor 1 to modulate suppression of translational termination. *Cell* **115:**319–331.

Park, H. S., Browning, K. S., Hohn, T., and Ryabova, L. A. (2004). Eucaryotic initiation factor 4B controls eIF3-mediated ribosomal entry of viral reinitiation factor. *EMBO J.* **23:**1381–1391.

Park, H. S., Himmelbach, A., Browning, K. S., Hohn, T., and Ryabova, L. A. (2001). A plant viral "reinitiation" factor interacts with the host translational machinery *Cell* **106:**723–733.

Park, Y. W., and Katze, M. G. (1995). Translational control by influenza virus. Identification of cis-acting sequences and trans-acting factors which may regulate selective viral mRNA translation. *J. Biol. Chem.* **270:**28433–28439.

Parkin, N. T., Chamorro, M., and Varmus, H. E. (1992). Human immunodeficiency virus type 1 gag-pol frameshifting is dependent on downstream mRNA secondary structure: Demonstration by expression *in vivo*. *J. Virol.* **66:**5147–5151.

Pause, A., Belsham, G. J., Gingras, A. C., Donze, O., Lin, T. A., Lawrence, J. C., Jr., and Sonenberg, N. (1994). Insulin-dependent stimulation of protein synthesis by phosphorylation of a regulator of 5′-cap function. *Nature* **371:**762–767.

Pelham, H. R. (1978). Leaky UAG termination codon in tobacco mosaic virus RNA. *Nature* **272:**469–471.

Pelletier, J., and Sonenberg, N. (1988). Internal initiation of translation of eukaryotic mRNA directed by a sequence derived from poliovirus RNA. *Nature* **334:**320–325.

Perera, R., Daijogo, S., Walter, B. L., Nguyen, J. H., and Semler, B. L. (2007). Cellular protein modification by poliovirus: The two faces of poly(rC)-binding protein. *J. Virol.* **81:**8919–8932.

Pestova, T. V., Kolupaeva, V. G., Lomakin, I. B., Pilipenko, E. V., Shatsky, I. N., Agol, V. I., and Hellen, C. U. (2001). Molecular mechanisms of translation initiation in eukaryotes. *Proc. Natl. Acad. Sci. USA* **98:**7029–7036.

Pestova, T. V., Lomakin, I. B., and Hellen, C. U. (2004). Position of the CrPV IRES on the 40S subunit and factor dependence of IRES/80S ribosome assembly. *EMBO Rep.* **5:**906–913.

Pestova, T. V., Lorsch, J. R., and Hellen, C. U. (2007). The mechanism of translation initiation in eukaryotes. *In* "Translational Control in Biology and Medicine" (M. B. Mathews, N. Sonenberg, and J. W. B. Hershey, eds.), pp. 87–128. CSHL Press, Cold Spring Harbor, NY.

Petty, I. T., and Jackson, A. O. (1990). Two forms of the major barley stripe mosaic virus nonstructural protein are synthesized *in vivo* from alternative initiation codons. *Virology* **177:**829–832.

Pfingsten, J. S., Costantino, D. A., and Kieft, J. S. (2006). Structural basis for ribosome recruitment and manipulation by a viral IRES RNA. *Science* **314:**1450–1454.

Pilipenko, E. V., Pestova, T. V., Kolupaeva, V. G., Khitrina, E. V., Poperechnaya, A. N., Agol, V. I., and Hellen, C. U. (2000). A cell cycle-dependent protein serves as a template-specific translation initiation factor. *Genes Dev.* **14:**2028–2045.

Piron, M., Vende, P., Cohen, J., and Poncet, D. (1998). Rotavirus RNA-binding protein NSP3 interacts with eIF4GI and evicts the poly(A) binding protein from eIF4F. *EMBO J.* **17:**5811–5821.

Pisarev, A. V., Chard, L. S., Kaku, Y., Johns, H. L., Shatsky, I. N., and Belsham, G. J. (2004). Functional and structural similarities between the internal ribosome entry sites of hepatitis C virus and porcine teschovirus, a picornavirus. *J. Virol.* **78:**4487–4497.

Pisarev, A. V., Hellen, C. U., and Pestova, T. V. (2007). Recycling of eukaryotic posttermination ribosomal complexes. *Cell* **131:**286–299.

Pisarev, A. V., Shirokikh, N. E., and Hellen, C. U. (2005). Translation initiation by factor-independent binding of eukaryotic ribosomes to internal ribosomal entry sites. *C. R. Biol.* **328:**589–605.

Pisareva, V. P., Pisarev, A. V., Hellen, C. U., Rodnina, M. V., and Pestova, T. V. (2006). Kinetic analysis of interaction of eukaryotic release factor 3 with guanine nucleotides. *J. Biol. Chem.* **281:**40224–40235.

Plant, E. P., and Dinman, J. D. (2006). Comparative study of the effects of heptameric slippery site composition on −1 frameshifting among different eukaryotic systems. *RNA* **12:**666–673.

Pooggin, M. M., Fütterer, J., Skryabin, K. G., and Hohn, T. (2001). Ribosome shunt is essential for infectivity of cauliflower mosaic virus. *Proc. Natl. Acad. Sci. USA* **98:**886–891.

Pöyry, T. A., Kaminski, A., Connell, E. J., Fraser, C. S., and Jackson, R. J. (2007). The mechanism of an exceptional case of reinitiation after translation of a long

ORF reveals why such events do not generally occur in mammalian mRNA translation. *Genes Dev.* **21:**3149–3162.

Pöyry, T. A., Kaminski, A., and Jackson, R. J. (2004). What determines whether mammalian ribosomes resume scanning after translation of a short upstream open reading frame? *Genes Dev.* **18:**62–75.

Prats, A. C., De Billy, G., Wang, P., and Darlix, J. L. (1989). CUG initiation codon used for the synthesis of a cell surface antigen coded by the murine leukemia virus. *J. Mol. Biol.* **205:**363–372.

Prüfer, D., Tacke, E., Schmitz, J., Kull, B., Kaufmann, A., and Rohde, W. (1992). Ribosomal frameshifting in plants: A novel signal directs the − 1 frameshift in the synthesis of the putative viral replicase of potato leafroll luteovirus. *EMBO J.* **11:**1111–1117.

Pyronnet, S., Imataka, H., Gingras, A. C., Fukunaga, R., Hunter, T., and Sonenberg, N. (1999). Human eukaryotic translation initiation factor 4G (eIF4G) recruits mnk1 to phosphorylate eIF4E. *EMBO J.* **18:**270–279.

Raaben, M., Groot Koerkamp, M. J., Rottier, P. J., and de Haan, C. A. (2007). Mouse hepatitis coronavirus replication induces host translational shutoff and mRNA decay, with concomitant formation of stress granules and processing bodies. *Cell Microbiol.* **9:**2218–2229.

Raju, R., Raju, L., Hacker, D., Garcin, D., Compans, R., and Kolakofsky, D. (1990). Nontemplated bases at the 5′ ends of Tacaribe virus mRNAs. *Virology* **174:**53–59.

Ramírez, B. C., Garcin, D., Calvert, L. A., Kolakofsky, D., and Haenni, A. L. (1995). Capped nonviral sequences at the 5′ end of the mRNAs of Rice hoja blanca virus RNA4. *J. Virol.* **69:**1951–1954.

Randall, R. E., and Goodbourn, S. (2008). Interferons and viruses: An interplay between induction, signalling, antiviral responses and virus countermeasures. *J. Gen. Virol.* **89:**1–47.

Rao, P., Yuan, W., and Krug, R. M. (2003). Crucial role of CA cleavage sites in the cap-snatching mechanism for initiating viral mRNA synthesis. *EMBO J.* **22:**1188–1198.

Rathjen, J. P., Karageorgos, L. E., Habili, N., Waterhouse, P. M., and Symons, R. H. (1994). Soybean dwarf luteovirus contains the third variant genome type in the luteovirus group. *Virology* **198:**671–679.

Raught, B., and Gingras, A.-C. (2007). Signaling to translation initiation. In "Translational Control in Biology and Medicine" (M. B. Mathews, N. Sonenberg, and J. W. B. Hershey, eds.), pp. 369–400. CSHL Press, Cold Spring Harbor, NY.

Rice, N. R., Stephens, R. M., Burny, A., and Gilden, R. V. (1985). The gag and pol genes of bovine leukemia virus: Nucleotide sequence and analysis. *Virology* **142:**357–377.

Rijnbrand, R., Bredenbeek, P. J., Haasnoot, P. C., Kieft, J. S., Spaan, W. J., and Lemon, S. M. (2001). The influence of downstream protein-coding sequence on internal ribosome entry on hepatitis C virus and other flavivirus RNAs. *RNA* **7:**585–597.

Rivera, C. I., and Lloyd, R. E. (2008). Modulation of enteroviral proteinase cleavage of poly(A)-binding protein (PABP) by conformation and PABP-associated factors. *Virology* **375:**59–72.

Riviere, C. J., and Rochon, D. M. (1990). Nucleotide sequence and genomic organization of melon necrotic spot virus. *J. Gen. Virol.* **71:**1887–1896.

Robaglia, C., and Caranta, C. (2006). Translation initiation factors: A weak link in plant RNA virus infection. *Trends Plant Sci.* **11:**40–45.

Robert, F., Kapp, L. D., Khan, S. N., Acker, M. G., Kolitz, S., Kazemi, S., Kaufman, R. J., Merrick, W. C., Koromilas, A. E., Lorsch, J. R., and Pelletier, J. (2006). Initiation of protein synthesis by hepatitis C virus is refractory to reduced eIF2·GTP·Met-tRNA(i)(Met) ternary complex availability. *Mol. Biol. Cell* **17:**4632–4644.

Rochon, D. M., and Tremaine, J. H. (1989). Complete nucleotide sequence of the cucumber necrosis virus genome. *Virology* **169:**251–259.

Rodríguez Pulido, M., Serrano, P., Sáiz, M., and Martínez-Salas, E. (2007). Foot-and-mouth disease virus infection induces proteolytic cleavage of PTB, eIF3a, b, and PABP RNA-binding proteins. *Virology* **364:**466–474.

Rohde, W., Gramstat, A., Schmitz, J., Tacke, E., and Prüfer, D. (1994). Plant viruses as model systems for the study of non-canonical translation mechanisms in higher plants. *J. Gen. Virol.* **75:**2141–2149.

Ryabov, E. V., Generozov, E. V., Kendall, T. L., Lommel, S. A., and Zavriev, S. K. (1994). Nucleotide sequence of carnation ringspot dianthovirus RNA-1. *J. Gen. Virol.* **75:**243–247.

Ryabova, L. A., Pooggin, M. M., and Hohn, T. (2002). Viral strategies of translation initiation: Ribosomal shunt and reinitiation. *Prog. Nucleic Acid Res. Mol. Biol.* **72:**1–39.

Ryabova, L. A., Pooggin, M. M., and Hohn, T. (2006). Translation reinitiation and leaky scanning in plant viruses. *Virus Res.* **119:**52–62.

Sadowy, E., Milner, M., and Haenni, A. L. (2001). Proteins attached to viral genomes are multifunctional. *Adv. Virus Res.* **57:**185–262.

Sanchez, A., Trappier, S. G., Mahy, B. W., Peters, C. J., and Nichol, S. T. (1996). The virion glycoproteins of Ebola viruses are encoded in two reading frames and are expressed through transcriptional editing. *Proc. Natl. Acad. Sci. USA* **93:**3602–3607.

Sanz, M. A., Castello, A., and Carrasco, L. (2007). Viral translation is coupled to transcription in Sindbis virus-infected cells. *J. Virol.* **81:**7061–7068.

Sasaki, J., and Nakashima, N. (1999). Translation initiation at the CUU codon is mediated by the internal ribosome entry site of an insect picorna-like virus *in vitro*. *J. Virol.* **73:**1219–1226.

Sasaki, J., and Nakashima, N. (2000). Methionine-independent initiation of translation in the capsid protein of an insect RNA virus. *Proc. Natl. Acad. Sci. USA* **97:**1512–1515.

Schalk, H. J., Matzeit, V., Schiller, B., Schell, J., and Gronenborn, B. (1989). Wheat dwarf virus, a geminivirus of graminaceous plants needs splicing for replication. *EMBO J.* **8:**359–364.

Scheets, K., and Redinbaugh, M. G. (2006). Infectious cDNA transcripts of Maize necrotic streak virus: Infectivity and translational characteristics. *Virology* **350:**171–183.

Scheper, G. C., and Proud, C. G. (2002). Does phosphorylation of the cap-binding protein eIF4E play a role in translation initiation? *Eur. J. Biochem.* **269:**5350–5359.

Schmitt, C., Balmori, E., Jonard, G., Richards, K. E., and Guilley, H. (1992). *In vitro* mutagenesis of biologically active transcripts of beet necrotic yellow vein virus RNA 2: Evidence that a domain of the 75-kDa readthrough protein is important for efficient virus assembly. *Proc. Natl. Acad. Sci. USA* **89:**5715–5719.

Schmitz, J., Prüfer, D., Rohde, W., and Tacke, E. (1996). Non-canonical translation mechanisms in plants: Efficient *in vitro* and in planta initiation at AUU codons of the tobacco mosaic virus enhancer sequence. *Nucleic Acids Res.* **24:**257–263.

Schneider, P. A., Kim, R., and Lipkin, W. I. (1997). Evidence for translation of the Borna disease virus G protein by leaky ribosomal scanning and ribosomal reinitiation. *J. Virol.* **71:**5614–5619.

Sciabica, K. S., Dai, Q. J., and Sandri-Goldin, R. M. (2003). ICP27 interacts with SRPK1 to mediate HSV splicing inhibition by altering SR protein phosphorylation. *EMBO J.* **22:**1608–1619.

Shen, R., and Miller, W. A. (2004). The 3′ untranslated region of tobacco necrosis virus RNA contains a barley yellow dwarf virus-like cap-independent translation element. *J. Virol.* **78:**4655–4664.

Shih, S. R., Nemeroff, M. E., and Krug, R. M. (1995). The choice of alternative 5′ splice sites in influenza virus M1 mRNA is regulated by the viral polymerase complex. *Proc. Natl. Acad. Sci. USA* **92:**6324–6328.

Shirako, Y. (1998). Non-AUG translation initiation in a plant RNA virus: A forty-amino-acid extension is added to the N terminus of the soil-borne wheat mosaic virus capsid protein. *J. Virol.* **72:**1677–1682.

Shirako, Y., and Wilson, T. M. (1993). Complete nucleotide sequence and organization of the bipartite RNA genome of soil-borne wheat mosaic virus. *Virology* **195:**16–32.

Siddell, S., Wege, H., Barthel, A., and ter Meulen, V. (1981). Intracellular protein synthesis and the *in vitro* translation of coronavirus JHM mRNA. *Adv. Exp. Med. Biol.* **142:**193–207.

Silvera, D., Gamarnik, A. V., and Andino, R. (1999). The N-terminal K homology domain of the poly(rC)-binding protein is a major determinant for binding to the poliovirus 5′-untranslated region and acts as an inhibitor of viral translation. *J. Biol. Chem.* **274:**38163–38170.

Simon-Buela, L., Guo, H. S., and Garcia, J. A. (1997). Cap-independent leaky scanning as the mechanism of translation initiation of a plant viral genomic RNA. *J. Gen. Virol.* **78:**2691–2699.

Siridechadilok, B., Fraser, C. S., Hall, R. J., Doudna, J. A., and Nogales, E. (2005). Structural roles for human translation factor eIF3 in initiation of protein synthesis. *Science* **310:**1513–1515.

Skotnicki, M. L., Mackenzie, A. M., Torronen, M., and Gibbs, A. J. (1993). The genomic sequence of cardamine chlorotic fleck carmovirus. *J. Gen. Virol.* **74:**1933–1937.

Skulachev, M. V., Ivanov, P. A., Karpova, O. V., Korpela, T., Rodionova, N. P., Dorokhov, Y. L., and Atabekov, J. G. (1999). Internal initiation of translation directed by the 5′-untranslated region of the tobamovirus subgenomic RNA I(2). *Virology* **263:**139–154.

Skuzeski, J. M., Nichols, L. M., Gesteland, R. F., and Atkins, J. F. (1991). The signal for a leaky UAG stop codon in several plant viruses includes the two downstream codons. *J. Mol. Biol.* **218:**365–373.

Slobin, L. I., and Moller, W. (1978). Purification and properties of an elongation factor functionally analogous to bacterial elongation factor Ts from embryos of *Artemia salina. Eur. J. Biochem.* **84:**69–77.

Sommergruber, W., Ahorn, H., Klump, H., Seipelt, J., Zoephel, A., Fessl, F., Krystek, E., Blaas, D., Kuechler, E., Liebig, H. D., and Skern, T. (1994). 2A proteinases of coxsackie- and rhinovirus cleave peptides derived from eIF-4γ via a common recognition motif. *Virology* **198:**741–745.

Sousa, C., Schmid, E. M., and Skern, T. (2006). Defining residues involved in human rhinovirus 2A proteinase substrate recognition. *FEBS Lett.* **580:**5713–5717.

Spahn, C. M., Jan, E., Mulder, A., Grassucci, R. A., Sarnow, P., and Frank, J. (2004). Cryo-EM visualization of a viral internal ribosome entry site bound to human ribosomes: The IRES functions as an RNA-based translation factor. *Cell* **118:**465–475.

Spahn, C. M., Kieft, J. S., Grassucci, R. A., Penczek, P. A., Zhou, K., Doudna, J. A., and Frank, J. (2001). Hepatitis C virus IRES RNA-induced changes in the conformation of the 40s ribosomal subunit. *Science* **291:**1959–1962.

Steward, M., Vipond, I. B., Millar, N. S., and Emmerson, P. T. (1993). RNA editing in Newcastle disease virus. *J. Gen. Virol.* **74:**2539–2547.

Stinski, M. F. (1977). Synthesis of proteins and glycoproteins in cells infected with human cytomegalovirus. *J. Virol.* **23:**751–767.

Stoltzfus, C. M., and Madsen, J. M. (2006). Role of viral splicing elements and cellular RNA binding proteins in regulation of HIV-1 alternative RNA splicing. *Curr. HIV Res.* **4:**43–55.

Strauss, E. G., and Strauss, J. H. (1991). RNA viruses: Genome structure and evolution. *Curr. Opin. Genet. Dev.* **1:**485–493.

Strauss, J. H., and Strauss, E. G. (1994). The alphaviruses: Gene expression, replication, and evolution. *Microbiol. Rev.* **58:**491–562.

Strong, R., and Belsham, G. J. (2004). Sequential modification of translation initiation factor eIF4GI by two different foot-and-mouth disease virus proteases within infected baby hamster kidney cells: Identification of the 3Cpro cleavage site. *J. Gen. Virol.* **85:**2953–2962.

Stuart, K. D., Weeks, R., Guilbride, L., and Myler, P. J. (1992). Molecular organization of Leishmania RNA virus 1. *Proc. Natl. Acad. Sci. USA* **89:**8596–8600.

Sudhakar, A., Ramachandran, A., Ghosh, S., Hasnain, S. E., Kaufman, R. J., and Ramaiah, K. V. (2000). Phosphorylation of serine 51 in initiation factor 2 alpha (eIF2 alpha) promotes complex formation between eIF2 alpha(P) and eIF2B and causes inhibition in the guanine nucleotide exchange activity of eIF2B. *Biochemistry* **39:**12929–12938.

Svitkin, Y. V., Gradi, A., Imataka, H., Morino, S., and Sonenberg, N. (1999). Eukaryotic initiation factor 4GII (eIF4GII), but not eIF4GI, cleavage correlates with inhibition of host cell protein synthesis after human rhinovirus infection. *J. Virol.* **73:**3467–3472.

Svitkin, Y. V., Herdy, B., Costa-Mattioli, M., Gingras, A. C., Raught, B., and Sonenberg, N. (2005). Eukaryotic translation initiation factor 4E availability controls the switch between cap-dependent and internal ribosomal entry site-mediated translation. *Mol. Cell. Biol.* **25:**10556–10565.

Tavazza, M., Lucioli, A., Calogero, A., Pay, A., and Tavazza, R. (1994). Nucleotide sequence, genomic organization and synthesis of infectious transcripts from a full-length clone of artichoke mottle crinkle virus. *J. Gen. Virol.* **75:**1515–1524.

Taylor, J. M. (2006). Hepatitis delta virus. *Virology* **344:**71–76.

ten Dam, E. B., Pleij, C. W., and Bosch, L. (1990). RNA pseudoknots: Translational frameshifting and readthrough on viral RNAs. *Virus Genes* **4:**121–136.

Terenin, I. M., Dmitriev, S. E., Andreev, D. E., Royall, E., Belsham, G. J., Roberts, L. O., and Shatsky, I. N. (2005). A cross-kingdom internal ribosome entry site reveals a simplified mode of internal ribosome entry. *Mol. Cell. Biol.* **25:**7879–7888.

Terenin, I. M., Dmitriev, S. E., Andreev, D. E., and Shatsky, I. N. (2008). Eukaryotic translation initiation machinery can operate in a bacterial-like mode without eIF2. *Nat. Struct. Mol. Biol.* **15:**836–841.

Thompson, S. R., Gulyas, K. D., and Sarnow, P. (2001). Internal initiation in *Saccharomyces cerevisiae* mediated by an initiator tRNA/eIF2-independent internal ribosome entry site element. *Proc. Natl. Acad. Sci. USA* **98:**12972–12977.

Tomonaga, K., Kobayashi, T., and Ikuta, K. (2002). Molecular and cellular biology of Borna disease virus infection. *Microbes Infect.* **4:**491–500.

Touriol, C., Bornes, S., Bonnal, S., Audigier, S., Prats, H., Prats, A. C., and Vagner, S. (2003). Generation of protein isoform diversity by alternative initiation of translation at non-AUG codons. *Biol. Cell* **95:**169–178.

Toyoda, H., Franco, D., Fujita, K., Paul, A. V., and Wimmer, E. (2007). Replication of poliovirus requires binding of the poly(rC) binding protein to the cloverleaf as well as to the adjacent C-rich spacer sequence between the cloverleaf and the internal ribosomal entry site. *J. Virol.* **81:**10017–10028.

Treder, K., Kneller, E. L., Allen, E. M., Wang, Z., Browning, K. S., and Miller, W. A. (2008). The 3′ cap-independent translation element of Barley yellow dwarf virus binds eIF4F via the eIF4G subunit to initiate translation. *RNA* **14:**134–147.

Uchida, N., Hoshino, S., Imataka, H., Sonenberg, N., and Katada, T. (2002). A novel role of the mammalian GSPT/eRF3 associating with poly(A)-binding protein in Cap/Poly(A)-dependent translation. *J. Biol. Chem.* **277:**50286–50292.

Valle, R. P., Drugeon, G., Devignes-Morch, M. D., Legocki, A. B., and Haenni, A. L. (1992). Codon context effect in virus translational readthrough. A study *in vitro* of the determinants of TMV and Mo-MuLV amber suppression. *FEBS Lett.* **306:**133–139.

van der Wilk, F., Dullemans, A. M., Verbeek, M., and Van den Heuvel, J. F. (1997). Nucleotide sequence and genomic organization of *Acyrthosiphon pisum* virus. *Virology* **238:**353–362.

van Eyll, O., and Michiels, T. (2002). Non-AUG-initiated internal translation of the L* protein of Theiler's virus and importance of this protein for viral persistence. *J. Virol.* **76:**10665–10673.

Veidt, I., Bouzoubaa, S. E., Leiser, R. M., Ziegler-Graff, V., Guilley, H., Richards, K., and Jonard, G. (1992). Synthesis of full-length transcripts of beet western yellows virus RNA: Messenger properties and biological activity in protoplasts. *Virology* **186:**192–200.

Veidt, I., Lot, H., Leiser, M., Scheidecker, D., Guilley, H., Richards, K., and Jonard, G. (1988). Nucleotide sequence of beet western yellows virus RNA. *Nucleic Acids Res.* **16:**9917–9932.

Vende, P., Piron, M., Castagne, N., and Poncet, D. (2000). Efficient translation of rotavirus mRNA requires simultaneous interaction of NSP3 with the eukaryotic translation initiation factor eIF4G and the mRNA 3′ end. *J. Virol.* **74:**7064–7071.

Ventoso, I., Blanco, R., Perales, C., and Carrasco, L. (2001). HIV-1 protease cleaves eukaryotic initiation factor 4G and inhibits cap-dependent translation. *Proc. Natl. Acad. Sci. USA* **98:**12966–12971.

Ventoso, I., Sanz, M. A., Molina, S., Berlanga, J. J., Carrasco, L., and Esteban, M. (2006). Translational resistance of late alphavirus mRNA to eIF2alpha phosphorylation: A strategy to overcome the antiviral effect of protein kinase PKR. *Genes Dev.* **20:**87–100.

Verchot, J., Angell, S. M., and Baulcombe, D. C. (1998). *In vivo* translation of the triple gene block of potato virus X requires two subgenomic mRNAs. *J. Virol.* **72:**8316–8320.

Versteeg, G. A., Slobodskaya, O., and Spaan, W. J. (2006). Transcriptional profiling of acute cytopathic murine hepatitis virus infection in fibroblast-like cells. *J. Gen. Virol.* **87:**1961–1975.

Verver, J., Le Gall, O., van Kammen, A., and Wellink, J. (1991). The sequence between nucleotides 161 and 512 of cowpea mosaic virus M RNA is able to support internal initiation of translation *in vitro*. *J. Gen. Virol.* **72:**2339–2345.

Vialat, P., and Bouloy, M. (1992). Germiston virus transcriptase requires active 40S ribosomal subunits and utilizes capped cellular RNAs. *J. Virol.* **66:**685–693.

Vidal, S., Curran, J., and Kolakofsky, D. (1990). A stuttering model for paramyxovirus P mRNA editing. *EMBO J.* **9:**2017–2022.

Volchkov, V. E., Becker, S., Volchkova, V. A., Ternovoj, V. A., Kotov, A. N., Netesov, S. V., and Klenk, H. D. (1995). GP mRNA of Ebola virus is edited by the Ebola virus polymerase and by T7 and vaccinia virus polymerases. *Virology* **214:**421–430.

Walsh, D., and Mohr, I. (2004). Phosphorylation of eIF4E by Mnk-1 enhances HSV-1 translation and replication in quiescent cells. *Genes Dev.* **18:**660–672.

Walsh, D., and Mohr, I. (2006). Assembly of an active translation initiation factor complex by a viral protein. *Genes Dev.* **20:**461–472.

Walsh, D., Perez, C., Notary, J., and Mohr, I. (2005). Regulation of the translation initiation factor eIF4F by multiple mechanisms in human cytomegalovirus-infected cells. *J. Virol.* **79:**8057–8064.

Walter, B. L., Parsley, T. B., Ehrenfeld, E., and Semler, B. L. (2002). Distinct poly (rC) binding protein KH domain determinants for poliovirus translation initiation and viral RNA replication. *J. Virol.* **76:**12008–12022.

Wang, J. Y., Chay, C., Gildow, F. E., and Gray, S. M. (1995). Readthrough protein associated with virions of barley yellow dwarf luteovirus and its potential role in regulating the efficiency of aphid transmission. *Virology* **206:**954–962.

Wang, S., Browning, K. S., and Miller, W. A. (1997). A viral sequence in the 3′-untranslated region mimics a 5′ cap in facilitating translation of uncapped mRNA. *EMBO J.* **16:**4107–4116.

Weissmann, C., Cattaneo, R., and Billeter, M. A. (1990). RNA editing. Sometimes an editor makes sense. *Nature* **343:**697–699.

White, K. A. (2002). The premature termination model: A possible third mechanism for subgenomic mRNA transcription in (+)-strand RNA viruses. *Virology* **304:**147–154.

White, K. A., Skuzeski, J. M., Li, W., Wei, N., and Morris, T. J. (1995). Immunodetection, expression strategy and complementation of turnip crinkle virus p28 and p88 replication components. *Virology* **211:**525–534.

Wilkie, G. S., Dickson, K. S., and Gray, N. K. (2003). Regulation of mRNA translation by 5′- and 3′-UTR-binding factors. *Trends Biochem. Sci.* **28:**182–188.

Willcocks, M. M., Carter, M. J., and Roberts, L. O. (2004). Cleavage of eukaryotic initiation factor eIF4G and inhibition of host-cell protein synthesis during feline calicivirus infection. *J. Gen. Virol.* **85:**1125–1130.

Wilson, J. E., Pestova, T. V., Hellen, C. U., and Sarnow, P. (2000). Initiation of protein synthesis from the A site of the ribosome. *Cell* **102:**511–520.

Wilson, J. E., Powell, M. J., Hoover, S. E., and Sarnow, P. (2000). Naturally occurring dicistronic cricket paralysis virus RNA is regulated by two internal ribosome entry sites. *Mol. Cell. Biol.* **20:**4990–4999.

Wittmann, S., Chatel, H., Fortin, M. G., and Laliberté, J. F. (1997). Interaction of the viral protein genome linked of turnip mosaic potyvirus with the translational eukaryotic initiation factor (iso) 4E of *Arabidopsis thaliana* using the yeast two-hybrid system. *Virology* **234:**84–92.

Xi, Q., Cuesta, R., and Schneider, R. J. (2004). Tethering of eIF4G to adenoviral mRNAs by viral 100k protein drives ribosome shunting. *Genes Dev.* **18:**1997–2009.

Xi, Q., Cuesta, R., and Schneider, R. J. (2005). Regulation of translation by ribosome shunting through phosphotyrosine-dependent coupling of adenovirus protein 100k to viral mRNAs. *J. Virol.* **79:**5676–5683.

Xiong, Z., Kim, K. H., Kendall, T. L., and Lommel, S. A. (1993). Synthesis of the putative red clover necrotic mosaic virus RNA polymerase by ribosomal frameshifting *in vitro. Virology* **193:**213–221.

Yamamiya, A., and Shirako, Y. (2000). Construction of full-length cDNA clones to soil-borne wheat mosaic virus RNA1 and RNA2, from which infectious RNAs are transcribed *in vitro*: Virion formation and systemic infection without expression of the N-terminal and C-terminal extensions to the capsid protein. *Virology* **277:**66–75.

Yamamoto, H., Nakashima, N., Ikeda, Y., and Uchiumi, T. (2007). Binding mode of the first aminoacyl-tRNA in translation initiation mediated by Plautia stali intestine virus internal ribosome entry site. *J. Biol. Chem.* **282:**7770–7776.

Yang, A. D., Barro, M., Gorziglia, M. I., and Patton, J. T. (2004). Translation enhancer in the 3′-untranslated region of rotavirus gene 6 mRNA promotes expression of the major capsid protein VP6. *Arch. Virol.* **149:**303–321.

Yang, W., and Hinnebusch, A. G. (1996). Identification of a regulatory subcomplex in the guanine nucleotide exchange factor eIF2B that mediates inhibition by phosphorylated eIF2. *Mol. Cell. Biol.* **16:**6603–6616.

Yeam, I., Cavatorta, J. R., Ripoll, D. R., Kang, B. C., and Jahn, M. M. (2007). Functional dissection of naturally occurring amino acid substitutions in eIF4E that confers recessive potyvirus resistance in plants. *Plant Cell* **19:**2913–2928.

Yilmaz, A., Bolinger, C., and Boris-Lawrie, K. (2006). Retrovirus translation initiation: Issues and hypotheses derived from study of HIV-1. *Curr. HIV Res.* **4:**131–139.

Yoshinaka, Y., Katoh, I., Copeland, T. D., and Oroszlan, S. (1985). Murine leukemia virus protease is encoded by the gag-pol gene and is synthesized through suppression of an amber termination codon. *Proc. Natl. Acad. Sci. USA* **82:**1618–1622.

Yu, Y., and Alwine, J. C. (2006). 19S late mRNAs of simian virus 40 have an internal ribosome entry site upstream of the virion structural protein 3 coding sequence. *J. Virol.* **80:**6553–6558.

Yu, Y., Kudchodkar, S. B., and Alwine, J. C. (2005). Effects of simian virus 40 large and small tumor antigens on mammalian target of rapamycin signaling: Small tumor antigen mediates hypophosphorylation of eIF4E-binding protein 1 late in infection. *J. Virol.* **79:**6882–6889.

Yueh, A., and Schneider, R. J. (2000). Translation by ribosome shunting on adenovirus and hsp70 mRNAs facilitated by complementarity to 18S rRNA. *Genes Dev.* **14:**414–421.

Zamora, M., Marissen, W. E., and Lloyd, R. E. (2002). Multiple eIF4GI-specific protease activities present in uninfected and poliovirus-infected cells. *J. Virol.* **76:**165–177.

Zhang, B., Morace, G., Gauss-Muller, V., and Kusov, Y. (2007). Poly(A) binding protein, C-terminally truncated by the hepatitis A virus proteinase 3C, inhibits viral translation. *Nucleic Acids Res.* **35:**5975–5984.

Zhou, H., and Jackson, A. O. (1996). Expression of the barley stripe mosaic virus RNA beta "triple gene block" *Virology* **216:**367–379.

Ziff, E. B. (1980). Transcription and RNA processing by the DNA tumour viruses. *Nature* **287:**491–499.

Ziff, E. B. (1985). Splicing in adenovirus and other animal viruses. *Int. Rev. Cytol.* **93:**327–358.

Zoll, W. L., Horton, L. E., Komar, A. A., Hensold, J. O., and Merrick, W. C. (2002). Characterization of mammalian eIF2A and identification of the yeast homolog. *J. Biol. Chem.* **277:**37079–37087.

Zvereva, S. D., Ivanov, P. A., Skulachev, M. V., Klyushin, A. G., Dorokhov, Y. L., and Atabekov, J. G. (2004). Evidence for contribution of an internal ribosome entry site to intercellular transport of a tobamovirus. *J. Gen. Virol.* **85:**1739–1744.

A

Abacavir, 31
Acridine compounds, 3
Acute human influenza A virus
    infection, 57
Acyclic guanosine analogs, 19
Acyclic nucleoside phosphonates, 6, 25–26
Acyclovir
    oral bioavailability, 19
    selectivity of antiviral action of, 18
    structure of, 18
Adamantanamine derivatives, 21
Adefovir dipivoxil
    as anti-HIV drug, 26
    for HBV infections treatment, 27
Amantadine, 20–21
Amino sulfonic acids, 3
Amprenavir, 37
Antigenic drift, 59
Anti-HIV compounds, 34
    amprenavir, 37
    atazanavir, 38
    darunavir, 39
    etravirine, 35
    fosamprenavir, 38
    indinavir, 36
    lopinavir, 37
    nelfinavir, 37
    nevirapine, 35
    peptidomimetics, 39
    rilpivirine, 35
    ritonavir, 36
    saquinavir, 36
    tipranavir, 38
Anti-influenza virus agent
    adamantanamine derivatives, 21
    amantadine, 20–21
    oseltamivir, 21–22
    zanamivir, 21–22
Antivaccinia virus activity, 4–5
Arabinosyladenine (Ara-A), 16–17
Arabinosylcytosine (Ara-C), 16
Arabinosyl nucleoside analogs, 16

Atazanavir, 38
Avian–human reassortant influenza
    virus, 59
Avian influenza H5N1 viruses
    human cases of, 60–61
    human-to-human transmission of, 74
Avian influenza viruses. *See also* Avian
    influenza H5N1 viruses
    gene knockout mice
        cytokines, 69
        host responses, 68
        kinetics of weight loss and
            mortality, 68
        Mx proteins, 69–70
    H7 subtype, 61
    interspecies transmission of, 58
    mammalian models to study, 82–83
    pandemic traits, 59
    pathogenesis, molecular basis of, 75
        avian HA subtypes, 77
        HA and NA, 77–78
        HA cleavage site, 76–77
        infection, 76
        molecular determinants of, 76
        NS1 protein, 81–82
        PB1-F2 protein, 81
        PB2 protein, 79–81
        polymerase proteins, 78–79
        viral proteins, 76
    transmissibility by respiratory
        droplets, 74–75
    transmission of H5/H7 viruses, 59
    tropism of, 70–71
    vaccination strategies, 83
Azidothymidine (AZT), 30

B

1-(β-D-ribofuranosyl)–2-bromo-5,6-
    dichlorobenzimidazolep
    (BDCRB), 11–12
Benzaldehyde thiosemicarbazone
    antibacterial chemotherapy, 4
    and isatin thiosemicarbazone, 5

Benzimidazole derivatives
    for CMV infection treatment
        BDCRB, 11–12
        maribavir, 12
    for influenza virus inhibition
        2,5-dimethylbenzimidazole, 9
        DRB, 9–10
        HBB, 10–11
Benzimidazoles, 8
Borna disease virus (BDV)
    protein translation, 108
    replication and transcription, 108
BVDU. *See* (E)–5-(2-bromovinyl)–2-
    deoxyuridine

                            C

CaMV pregenomic 35S mRNA, organization
    of, 125
Cap-dependent translation
    in eukaryotic viruses, 112
    in monocistronic mRNAs, 111
CCR5 antagonists, 40
Cell factor modification, 128
    eIF2α phosphorylation
        kinases role in, 130
        in MHV, 131
        in SFV, 130–131
        in virus-infected cells, 129–130
    eIF4E and 4E-BP, 131
        dephosphorylation, 132–134
        phosphorylation, 134
    eIF4G
        cleavage, 134–135
        phosphorylation, 135–136
    PABP, 136
    PBCP2 cleavage, 137
Cellular RNA splicing machinery, 108
Cidofovir
    for smallpox treatment, 6
    structure of, 7
Circularization
    binding of initiation factor, 112
    direct base pairing, 114
    models of, 114
    RNA–protein and protein–protein
        interactions, 112
Closed-loop model
    binding of eIF4E to 5′ cap, 112
    Poliovirus RNA, 114
    for translation, 112, 114
Coreceptor inhibitors, 40
Cricket paralysis virus

eIF2α phosphorylation, 130
IRESs in, 118, 130
CRIs. *See* Coreceptor inhibitors
CrPV. *See* Cricket paralysis virus

                            D

Darunavir, 39
3-Deazaneplanocin, 24
Delavirdine, 35
DHPA. *See* (S)–9-(2,3-Dihydroxypropyl)
    adenine
2,6-Diaminopurine, 4
5,6-Dichloro-1-β-D-
    ribofuranosylbenzimidazole, 9–10
Dicistronic constructs, ORF of, 125–126
Didanosine, 31
(S)–9-(2,3-Dihydroxypropyl)adenine, 23–24
2,5-Dimethylbenzimidazole, 9
5,6-Dimethylbenzimidazole, 9
DRB. *See* 5,6-Dichloro-1-β-D-
    ribofuranosylbenzimidazole
Duviragel®. *See* (S)–9-(2,3-
    Dihydroxypropyl)adenine

                            E

(E)–5-(2-bromovinyl)–2-
    deoxyuridine, 15–16
Editing. *See* RNA editing
eEF-1 and eEF-2, 140
Efavirenz, 35
eIF2α phosphorylation
    kinases role in, 130
    in MHV, 131
    in SFV, 130–131
    in virus-infected cells, 129–130
eIF4E
    function in potyvirus cycle, 128–129
    variation in, 128
    and 4E-BP, 131
        dephosphorylation, 132–134
        phosphorylation, 134
eIF4G
    cleavage, 134–135
    phosphorylation, 135–136
eIFiso4E, variation in, 128
E627K substitution, 80
Elongation factors, modification of, 140
EMCV IRES-mediated initiation of
    translation, 117
Emtricitabine, 32
Enfuvirtide, 40–41
Etravirine, 35

F

Famciclovir, 20
Ferret model
  for H5N1 virus pathogenesis
    inflammation in lungs and brains, 73
    pathogenicity phenotype, 72
  for human influenza A virus
    pathogenesis, 71
  for H7 virus pathogenesis, 73
  VN/1203 virus in, 82
FI. *See* Fusion inhibitor
FMDV IRES-mediated translation, 117
Fosamprenavir, 38
Frameshifting
  elongation of translation by, 138–140
  RNA signal, 137
Fusion inhibitor, 40

G

gag–pol precursor proteins, 39
Ganciclovir
  anti-CMV effects of, 12
  structure, 20
Glycoproteins (GPs) of Ebola
  virus, 107–108

H

Hard gel capsules, 36
HBB. *See* 2-(1-hydroxybenzyl)-
  benzimidazole
HCV IRES, 118
HCV-like IRESs, 118
Hemagglutinin cleavage site
  of H5 and H7 subtype, 76
  multibasic amino acid insertion within, 77
Hemagglutinin (HA), 57
  and NA activity, 77
Hepatitis C treatment, interferon and
  ribavirin for, 22–23
Hepatitis C virus
  initiation at IRES of, 117
  IRES stem–loop structure, 118–119
Hepatitis delta virus (HDV)
  genome, 104
  propagation, 106
  replication, RNA editing of antigenomic
    RNA, 104–105
HEPT derivatives, 32–34
Herpetic keratitis treatment
  with IDU and TFT, 15
  with 5-iodouridine, 14

H7 influenza A virus pathogenesis, mouse
  model for
  H7N1 viruses, 68
  HPAI H7N7 virus A/NL/219/03
    (NL/219), 67–68
HIV infections treatment
  fusion inhibitor (FI) for, 40
  TDF for, 28–29
HIV protease inhibitors, 39
H5N1 HA protein, 77
H5N1 human cases
  clinical symptoms, 61
H5N1 human virus infection, 60
H5N1 virus
  NA stalk deletions in, 77–78
  pathogenesis of. *See* H5N1 virus
    pathogenesis
  systemic spread of, 66
  virulence in humans, 74
H7N7 viruses
  amino acid substitution in PB2 protein
    in, 80
  NL/219, 70
  virulent, 82
H5N1 virus pathogenesis
  ferret model for
    inflammation in lungs and brains, 73
    pathogenicity phenotype, 72
  mouse model for
    HK/483-infected lung, 65–66
    HK/486-infected mice, 66
    infection with SP/83 viruses, 67
    infection with Thai/ 16 virus, 67
    innate immune/inflammatory
      responses, 65
    pathogenicity phenotypes, 65, 67
Host translational shutoff, 128
HPAI viruses, 60
  direct transmission to humans, 60
  of H5 and H7 subtype, 77
  HPAI A viruses, human infections and
    fatalities by, 83–84
  HPAI H7N3 outbreak, 61–62
  HPAI H7N7 virus
    HPAI H7N7 L/219 virus, 75
    NL/33, 78
    outbreak of, 61
  HPAI H5N1 viruses
    amino acid substitution in PB2
      protein in, 79–80
    E627K substitution in, 80
    human cases of, 60–61, 66–67

HPAI viruses (*cont.*)
    pathogenicity, 82
    poultry outbreaks due to, 67
    viral attachment of, 58
    virulent in mammalian models, 79
HPMPA. *See* (S)–9-(3-Hydroxy-2-
    phosphonylmethoxypropyl) adenine
Human CMV infection treatment
    BDCRB for, 11–12
    maribavir for, 12
Human H3N2 virus
    polymerase complex in, 79
    transmission of, 74
Human influenza A virus pathogenesis
    ferret model for, 71
    mouse model for, 63
        disadvantage of, 64
        molecular changes, 64–65
        WSN virus, 65
Human influenza viruses. *See also* Human
    influenza A virus pathogenesis
    hemagglutinin cleavage site, 76
    H1–H3 subtypes binding to bind sialic
        acid, 58
Human parainfluenza virus 1, initiation
    codon in, 119
H7 viruses
    human-to-human transmission of, 75
    pathogenesis, ferret model for, 73
2-(1-Hydroxybenzyl)-benzimidazole, 10–11
(S)–9-(3-Hydroxy-2-
    phosphonylmethoxypropyl)
    adenine, 25–26
    sister compound of, 26–27

I

IGRs. *See* Intergenic regions
Indinavir, 36
Influenza antiviral drugs, 83
Influenza A virus
    multiplication, inhibition of
        5,6-dimethylbenzimidazole for, 9
        DRB, 9–10, 20
    polymerase proteins of, 78–79
    RNA splicing in, 110
    subtypes, 57–58
    winter epidemics of, 56
Influenza B virus
    multiplication, inhibition of
        5,6-dimethylbenzimidazole for, 9
        DRB, 9–10, 20
    RNA splicing in, 110

Influenza, prevalence of, 57
Influenza viruses. *See also* Influenza A virus;
    Influenza B virus
    asymptomatic infection, 58
    binding to sialic acids, 58–59
    orthomyxoviridae, 109
    pandemic, 59
    pathogenesis
        ferret model for. *See* Ferret model
        mouse model for. *See* Mouse model
        virus–host interactions, 82
    research, mouse and ferret models
        for, 62–63
    RNA genome, 109
Integrase, 40
Interferon, 22
Intergenic regions, 103
5-Iodo-2′-deoxyuridine, 13
5-Iodouridine
    antiviral potential, 13–14
    deoxyribosyl counterpart, 13
    herpetic keratitis treatment by, 14–15
    and TFT, 14–15
IRES-directed initiation, 116
    cistron expression, 117
    rates of, 118
    sequence elements involved in
        eIF4F, 118
        encephalomyocarditis virus, 117
        FMDV, 117
        hepatovirus RNAs, 117–118
        mRNA-binding proteins, 117
        plant virus, 119
    TEs and, 119
Isatin thiosemicarbazone, 5
IUDR. *See* 5-Iodo-2′-deoxyuridine
IUR. *See* 5-Iodouridine

L

Lamivudine, 31
Leaky scanning mechanism
    in-frame initiation, 123
    overlapping ORFs
        in carlaviruses and
            potexviruses, 124
        codon context-dependent, 123
        for polycistronic RNAs expression, 122
Low pathogenicity avian influenza (LPAI)
    viruses, 60
LPAI H7N2 virus, 61
LPAI H7N3 virus, 62
LPAI H7 virus virulence, 77

M

Maraviroc, 41
Maribavir, 12
Methisazone, 5
Moloney murine leukemia virus
    (MoMLV), 122
Mouse model
    for H7 influenza A virus pathogenesis
        H7N1 viruses, 68
        HPAI H7N7 virus A/NL/219/03
            (NL/219), 67–68
    for H5N1 virus pathogenesis
        HK/483-infected lung, 65–66
        HK/486-infected mice, 66
        infection with SP/83 viruses, 67
        infection with Thai/16 virus, 67
        innate immune/inflammatory
            responses, 65
        pathogenicity phenotypes, 65, 67
    for human influenza A virus
        pathogenesis, 63
        disadvantage of, 64
        molecular changes, 64–65
        WSN virus, 65
mRNA circularization. *See* Circularization

N

NA. *See* Neuraminidase
NA stalk deletions, 77–78
Nelfinavir, 37
Neplanocin, 25
Neuraminidase, 57
    stalk deletions, 77–78
    subtypes, 58
    and virus pathogenicity, 78
Nevirapine, 35
N-methylisatin β-thiosemicarbazone, 5
NNRTIs. *See* Non-nucleoside reverse
    transcriptase inhibitors
Non-AUG initiation codons
    AUG codons, 119
    CUG codon, 122
    as initiators of protein synthesis, 120–121
    and methionine-independent translation
        initiation, 122
Non-nucleoside reverse transcriptase
    inhibitors
    HEPT, 32–33
    TIBO derivatives, 33–34
NRTIs. *See* Nucleoside reverse transcriptase
    inhibitors

NS1 protein
    antiviral activity of, 81
    deletion of, 81–82
    role in virus pathogenesis, 82
Nucleoside reverse transcriptase inhibitors
    with AZT, 32
    examples, 31
    mode of action of, 30
Nucleotide addition, RNA editing by
    GPs of Ebola virus, 107–108
    ORFs, 106
    paramyxoviruses, 106
    P/C mRNA, 107
Nucleotide modification, RNA editing by
    antigenome, 104
    HDV propagation, 106
    HDV replication, 104–105

O

Oral prodrug, 19
ORFs. *See* Overlapping open reading frames
Orthomyxoviridae, 57, 109
Orthopox, oral treatment of, 6–7
Orthopoxvirus replication inhibition
    compounds
    4-anilinoquinazoline CI-1033, 7
    ST-246, 8
Oseltamivir, 21–22
Overlapping open reading frames, 103
    in antigenome, 108
    in complementary antigenomic RNA, 106
    in RNA genomes of eukaryotic
        viruses, 110

P

PAA. *See* Phosphonoacetic acid
PABP. *See* Poly(A)-binding protein
Pandemic influenza, 56
Paramyxovirinae, 106
Paramyxoviruses, RNA genome of, 106
PBCP2 cleavage, 137
PB1-F2 protein, 81
PB2 protein
    amino acid substitution in, 79–80
    D701N mutation in, 80
    role in viral mRNA synthesis, 79
Peanut clump virus (PCV), 124
Penciclovir, 20
Penicillium rubrum, 3
Peptidomimetics, 39
PFA. *See* Phosphonoformic acid

Phosphonoacetic acid, 25–26
Phosphonoformic acid, 25–26
9-(2-Phosphonylmethoxyethyl)-
    adenine, 26–27
(R)–9-(2-Phosphonylmethoxypropyl)
    adenine, 28
PIs. *See* HIV protease inhibitors
Plasmid-based reverse genetics
    techniques, 76
PMEA. *See* 9-(2-
    Phosphonylmethoxyethyl)-adenine
PMPA See,
    (R)–9–(2–Phosphonylmethoxypropyl)
    adenine
Poly(A)-binding protein
    cleavage, 136
    closed-loop complex formation and, 112
    eIF4G binding with, 112
    and PCBP2, 114
    substitution, 136
Polymerase complex
    of influenza A viruses, 78–79
    role in virulence of avian influenza
        viruses, 79

                        R

Raltegravir, 41
Readthrough
    cis elements on mRNA for, 144
    in mouse cells infected with
        MLV, 144–145
    regulating termination of translation
        by, 142–144
    of termination codons, 141
    in TMV, 144
Receptor-binding preference, 58
Reinitiation of translation, 125
Release factors, binding of, 146
Ribavirin
    action mechanism of, 23
    for hepatitis C treatment, 22
    for severe respiratory tract infections, 23
Ribosome shunting mechanism
    in polycistronic mRNA of CaMV, 127
    in polycistronic P/C mRNA of Sendai
        virus, 128
Rilpivirine, 35
Ritonavir, 36
RNA editing
    definition, 104
    nucleotide addition
        GPs of Ebola virus, 107–108

ORFs, 106
paramyxoviruses, 106
P/C mRNA, 107
nucleotide modification
    antigenome, 104
    HDV propagation, 106
    HDV replication, 104–105
Rotaviruses, 115

                        S

SAH hydrolase inhibitors, 24
Saquinavir, 36
Seasonal influenza, 21
    outbreaks of, 57
    virus replication in, 58
Sendai virus
    P/C mRNA expression, 107
    P mRNA, 119
Smallpox, antivaccinial gamma-globulin in
        prophylaxis of, 5–6
Soft gelatin capsules, 36
Splicing
    definition, 108
    in influenza A virus and Influenza B
        virus, 110
    of introns I and II, 108
    S1–S3, transcription initiation sites, 109
ST-246, 8
Stavudine, 31
Subgenomic RNA synthesis, 110–111
5-Substituted 2′-deoxyuridine
    derivatives, 15–16
Suppressor tRNAs
    of UAG/UAA codons, 145
    of UGA codons, 145–146
Suramin
    antiviral drug, 29
    structure of, 30

                        T

TAV protein, 125
TDF. *See* Tenofovir disoproxil fumarate
Tenofovir disoproxil fumarate
    with emtricitabine and efavirenz, drug
        combination of, 29
    HIV infections treatment using, 28
Termination of translation
    readthrough, 141–145
    release factor binding for, 146
    suppressor tRNAs role in, 145–146
    termination codon recognition in, 141

TFT. *See* Trifluorothymidine
Thiosemicarbazones
  benzaldehyde, 4
  test against vaccinia virus, 4–5
TIBO derivatives, 33–34
Tipranavir, 38
Translation
  elongation
    elongation factors modification, 140
    frameshift, 137–140
  initiation
    Cap-dependent, 111–112
    cleavage of PABP for, 136–137
    cleavage of PBCP2 for, 137
    closed-loop model, 112–115
    dephosphorylation of eIF4E and of
      4E-BP1 for, 132–134
    IRES-directed, 116–119
    leaky scanning mechanism for, 122–124
    modification of eIF4E and 4E-BP
      for, 131–132
    modification of eIF4G for, 134–136
    on non-AUG initiation codons, 119–122
    phosphorylation of eIF2α for, 129–131
    phosphorylation of eIF4E and of 4E-
      BP1 for, 134
    reinitiation of translation of
      downstream ORFs for, 125–127
    ribosome shunting
      mechanism, 127–128
    substitution of PABP for, 136
    VPg and, 115–116
  regulation strategies prior to
    editing. *See* RNA editing
    sgRNA production, 110–111
    splicing, 108–110
  termination

    readthrough, 141–145
    release factor binding for, 146
    suppressor tRNAs role in, 145–146
    termination codon recognition in, 141
Trifluorothymidine, 14
Turnip yellow mosaic virus, 124

U

UAG/UAA codons, suppressor of, 145–146
UGA codons, suppressor of, 145–146
3′ UTR of mRNA
    initiation factors and, 115
    RNA–RNA interaction and, 115
    for translation, 114

V

Vaccination strategies for avian influenza
    viruses, 83
Valganciclovir, 20
Viral proteins, 76
Viral RdRp, endonuclease activity of, 110
Viral 5′ UTRs, 116
Viruses, regulation strategies prior to
    translation
    editing. *See* RNA editing
    sgRNA production, 110–111
    splicing, 108–110
VPg and initiation, 115–116
VP-Pro-induced degradation
    of RNAs, 128

Z

Zalcitabine, 31
Zanamivir, 21–22
Zidovudine, 31